Praise for *Gut Feelings*

Named as the 2007 Science Book of the Year by *Bild der Wissenschaft* journal and awarded the 2007 Handelszeitung Business Book Prize

"Gladwell drew heavily on Gigerenzer's research. But Gigerenzer goes a step further by explaining just why our gut instincts are so often right. Intuition, it seems, is not some sort of mystical chemical reaction but a neurologically based behavior that evolved to ensure that we humans respond quickly when faced with a dilemma." —*Business Week*

"Memorable. Clever. Gerd Gigerenzer, director of the Max Planck Institute for Human Development in Berlin, locates specific strategies that the unconscious mind uses to solve problems. These are not impulsive or capricious responses, but evolved methods that lead to superior choices. In short chapters, using vivid examples and ordinary language, Gigerenzer explains how an outfielder catches a fly ball not by complex calculations but by unconsciously adjusting his running speed so that the angle of his gaze at the ball remains constant. In problem solving, having too much information is often as harmful as having too little; having just enough information works best." —*The Boston Globe*

"Logic be damned! See how doctors really make tough diagnoses, how police spot drug couriers, and more. A convincing argument." —*Men's Health*

"A pleasing, edifying tout of territory that has long been dark and unobserved." —*Kirkus Reviews*

"Trust your hunches, for intuition does have an underlying rationale, according to this accessible account. [Gigerenzer] succeeds in converting a specialized topic into a conduit for greater self-awareness among his readers." —*Booklist*

"Readers who found Malcolm Gladwell's *Blink* stronger on anecdote than analysis will welcome this incisive study by a psychologist whose research provided one of the bases for Mr. Gladwell's bestseller." —*The New York Observer*

"Without intuition we'd drown in a sea of data points. Drawing on seven years of research, Gigerenzer argues counterintuitively that hunches and facts should be equally valid reasons for a search warrant and that the game show contestant who knows the least is often the most likely to nail the million dollar question." —*Time*

ABOUT THE AUTHOR

Gerd Gigerenzer is the director of the Center for Adaptive Behavior and Cognition at the Max Planck Institute for Human Development in Berlin, Germany. He has won numerous prizes, including the 1991 AAAS Prize for Behavioral Science Research and the 2002 German Science Book of the Year Prize. He has been professor of psychology at the University of Chicago and the John M. Olin Distinguished Visiting Professor at the School of Law, University of Virginia.

Gut Feelings

THE INTELLIGENCE
OF THE UNCONSCIOUS

GERD GIGERENZER

PENGUIN BOOKS

PENGUIN BOOKS

Published by the Penguin Group

Penguin Group (USA) Inc., 375 Hudson Street, New York, New York 10014, U.S.A.

Penguin Group (Canada), 90 Eglinton Avenue East, Suite 700, Toronto,
Ontario, Canada M4P 2Y3 (a division of Pearson Penguin Canada Inc.)

Penguin Books Ltd, 80 Strand, London WC2R 0RL, England

Penguin Ireland, 25 St Stephen's Green, Dublin 2, Ireland (a division of Penguin Books Ltd)

Penguin Group (Australia), 250 Camberwell Road, Camberwell,
Victoria 3124, Australia (a division of Pearson Australia Group Pty Ltd)

Penguin Books India Pvt Ltd, 11 Community Centre,
Panchsheel Park, New Delhi – 110 017, India

Penguin Group (NZ), 67 Apollo Drive, Rosedale, North Shore 0632,
New Zealand (a division of Pearson New Zealand Ltd)

Penguin Books (South Africa) (Pty) Ltd, 24 Sturdee Avenue,
Rosebank, Johannesburg 2196, South Africa

Penguin Books Ltd, Registered Offices:
80 Strand, London WC2R 0RL, England

First published in the United States of America by Viking Penguin,
a member of Penguin Group (USA) Inc. 2007
Published in Penguin Books 2008

10 9 8 7 6 5 4 3 2 1

Illustrations by Jurgen Rossbach

THE LIBRARY OF CONGRESS HAS CATALOGED THE HARDCOVER EDITION AS FOLLOWS:
Gigerenzer, Gerd.
    Gut feelings : the intelligence of the unconscious / Gerd Gigerenzer.
        p. cm.
    Includes bibliographical references and index.
    ISBN 978-0-670-03863-3 (hc.)
    ISBN 978-0-14-311376-8 (pbk.)
        1. Intuition.   I. Title.
    BF315.5.G54 2007
    153. 4'4—dc22   2006052810

Printed in the United States of America
Set in Dante MT with Frutiger
Designed by Daniel Lagin

In affectionate memory of my mother and her courage, humor, and patience

CONTENTS

We know more than we can tell.

—Michael Polanyi

Part 1 ⋮

UNCONSCIOUS INTELLIGENCE

The heart has its reasons of which reason knows nothing.

—Blaise Pascal

1 ⋮ GUT FEELINGS

We think of intelligence as a deliberate, conscious activity guided by the laws of logic. Yet much of our mental life is unconscious, based on processes alien to logic: gut feelings, or intuitions. We have intuitions about sports, friends, which toothpaste to buy, and other dangerous things. We fall in love, and we sense that the Dow Jones will go up. This book asks: Where do these feelings come from? How do we know?

Can following your gut feelings lead to some of the best decisions? It seems naive, even ludicrous, to think so. For decades, books on rational decision making, as well as consulting firms, have preached "look before you leap" and "analyze before you act." Pay attention. Be reflective, deliberate, and analytic. Survey all alternatives, list all pros and cons, and carefully weigh their utilities by their probabilities, preferably with the aid of a fancy statistical software package. Yet this scheme does not describe how actual people—including the authors of these books—reason. A professor from Columbia University was struggling over whether to accept an offer from a rival university or to stay. His colleague took him aside and said, "Just maximize your expected utility—you always write about doing this." Exasperated, the professor responded, "Come on, this is serious."

From economists to psychologists to John Q. Public, most people readily accept that the ideal of perfect beings with boundless knowledge and eternal time is unrealistic. Yet, they argue, without these bounds, and with more logic, we would make superior choices: we may not consider every angle, but we should. This message is not what you are going to read in the following pages.

With this book, I invite you on a journey into a largely unknown land of rationality, populated by people just like us, who are partially ignorant, whose time is limited and whose future is uncertain. This land is not one many scholars write about. They prefer to describe a land where the sun of enlightenment shines down in beams of logic and probability, whereas the land we are visiting is shrouded in a mist of dim uncertainty. In my story, what seem to be "limitations" of the mind can actually be its strengths. How the mind adapts and economizes by relying on the unconscious, on rules of thumb, and on evolved capacities is what *Gut Feelings* is about. The laws in the real world are puzzlingly different from those in the logical, idealized world. More information, even more thinking, is not always better, and less can be more. Ready for a glimpse?

THE HEART'S CHOICE

A close friend of mine (call him Harry) once found himself with two girlfriends, both of whom he loved, desired, and admired. Two, however, were one too many. Confused by contradicting emotions and unable to make up his mind, he recalled what Benjamin Franklin had once advised a nephew in a similar situation:

April 8, 1779

If you doubt, set down all the Reasons, pro and con, in opposite Columns on a Sheet of Paper, and when you have considered

them two or three Days, perform an Operation similar to that
in some questions of Algebra; observe what Reasons or Motives
in each Column are equal in weight, one to one, one to two,
two to three, or the like, and when you have struck out from
both Sides all the Equalities, you will see in which column remains the Balance. [. . .] This kind of *Moral Algebra* I have often practiced in important and dubious Concerns, and tho' it
cannot be mathematically exact, I have found it extreamly useful. By the way, if you do not learn it, I apprehend you will
never be married.

I am ever your affectionate Uncle,

B. FRANKLIN[1]

Harry was greatly relieved that a logical formula existed to
solve his conflict. So he took his time, wrote down all the important reasons he could think of, weighed them carefully, and went
through the calculation. When he saw the result, something unexpected happened. An inner voice told him that it wasn't right. And
for the first time, Harry realized that his heart had already
decided—against the calculation and in favor of the other girl. The
calculus helped to find the solution, but not because of its logic. It
brought an unconscious decision to his awareness, based on reasons obscure to him.

Thankful for the sudden solution but puzzled by the process,
Harry asked himself how it was possible to make unconscious
choices in contradiction to one's deliberate reasoning. He was
not the first to learn that reasoning can conflict with what we
call intuition. Social psychologist Timothy Wilson and his colleagues once offered posters to two groups of women as a thank-
you present for participating in an experiment.[2] In one group,
each woman simply picked her favorite poster out of a selection
of five; in a second group, each was asked to describe her reasons

for liking or disliking each poster before choosing one. Interestingly, the two groups tended to take different posters home. Four weeks later, they were all asked how much they enjoyed their present. Those who had given reasons were less satisfied and regretted their choice more than those who had not given any. Here and in similar experiments, deliberate thinking about reasons seems to lead to decisions that make us less happy, just as consciously thinking about how to ride a bike or put on a spontaneous smile is not always better than its automatic version. The unconscious parts of our mind can decide without us—the conscious self—knowing its reasons, or, as in Harry's case, without being aware that a decision has been made in the first place.

But isn't the capability for self-reflection uniquely human and therefore uniformly beneficial? After all, doesn't thinking about thinking define human nature? Freud used self-introspection as a therapeutic method, and decision consultants employ modern versions of Franklin's moral algebra as a rational tool. But the evidence suggests that weighing pros and cons does not generally make us happy. In one study, people were asked about various everyday activities such as how to decide which TV programs to watch in the evening and what to buy in a department store. Did they survey all channels, using the remote control to flip back and forth through all TV stations, constantly checking for a better program? Or did they quickly stop searching and watch a good-enough program? People who reported exhaustive search in shopping and leisure were called *maximizers*, because they tried hard to get the best. Those who engaged in a limited search and settled quickly with the first alternative that was satisfactory or "good enough" were called *satisficers*.[3] Satisficers were reported to be more optimistic and have higher self-esteem and life satisfaction, whereas maximizers excelled in depression, perfectionism, regret, and self-blame.

A BENEFICIAL DEGREE OF IGNORANCE

Imagine you are a contestant in a TV game show. You have out-witted all other competitors and eagerly await the $1 million question. Here it comes:

> Which city has the larger population,
> Detroit or Milwaukee?

Ouch, you have never been good in geography. The clock is ticking away. Except for the odd Trivial Pursuit addict, few people know the answer for sure. There is no way to logically deduce the correct answer; you have to use what you know and make your best guess. You might recall that Detroit is an industrial city, the birthplace of Motown and the American automobile industry. However, Milwaukee is also an industrial city, known for its breweries, and you might recall Ella Fitzgerald singing about her squawky cousin from there. What can you conclude from this?

Daniel Goldstein and I asked a class of American college students, and they were divided—some 40 percent voted for Milwaukee, the others for Detroit. Next we tested an equivalent class of German students. Virtually everyone gave the right answer: Detroit. One might conclude that the Germans were smarter, or at least knew more about American geography. Yet the opposite was the case. They knew very little about Detroit, and many of them had not even heard of Milwaukee. These Germans had to rely on their intuition rather than on good reasons. What is the secret of this striking intuition?

The answer is surprisingly simple. The Germans used a rule of thumb known as the *recognition heuristic*:[4]

> If you recognize the name of one city but not that of the other, then infer that the recognized city has the larger population.

The American students could *not* use this rule of thumb because they had heard of both cities. They knew too much. Myriad facts muddled their judgment and prevented them from finding the right answer. A beneficial degree of ignorance can be valuable, although relying on name recognition is of course not foolproof. For instance, Japanese tourists will likely falsely infer that Heidelberg is larger than Bielefeld, not having heard of the latter. Nonetheless, this rule gets the answer right in most cases and does so better than a considerable amount of knowledge does.

The recognition heuristic is not only helpful for winning a million dollars. People tend to rely on it, for example, when purchasing products whose brand name they recognize. Corporations in turn exploit consumers' heuristics, or rules of thumb, by investing in uninformative advertisement whose sole purpose is to increase brand-name recognition. The instinct to go with what you know has survival value in the natural world. Recall Dr. Seuss's famous menu of green eggs and ham; wouldn't you opt for the less exotic variety? By simply relying on familiar foods, you obtain the needed calories without wasting time or tempting fate by learning first-hand whether the eggs and ham are inedible or even toxic.

WINNING WITHOUT THINKING

How does a player catch a fly ball in baseball or cricket? If you ask a professional player, he'll likely stare blankly at you and say he'd never thought about it. A friend of mine named Phil played baseball for the local team. His coach scolded him repeatedly for being lazy, because Phil sometimes trotted over, as others did, toward the point where the ball came down. The coach thought that Phil took undue risks and insisted that he run as fast as he could in order to make any necessary last-second corrections. Phil found himself in a dilemma. When he and his teammates tried to avoid the coach's fury by running at top speed, they missed the ball more often. What was going wrong? Phil had played as an outfielder for years and had never understood how he caught the ball. His coach, in contrast, had a theory: He believed that players intuitively calculate the ball's trajectory, and that the best strategy is to run as fast as possible to the spot where the ball will hit the ground. How else could it work?

Phil's coach is not the only one who thinks of computing trajectories. In *The Selfish Gene*, biologist Richard Dawkins writes,

> When a man throws a ball high in the air and catches it again, he behaves as if he had solved a set of differential equations in predicting the trajectory of the ball. He may neither know nor care what a differential equation is, but this does not affect his skill with the ball. At some subconscious level, something functionally equivalent to the mathematical calculations is going on.[5]

Computing the trajectory of a ball is not a simple feat. Theoretically, balls have parabolic trajectories. In order to select the right parabola, the player's brain would have to estimate the ball's initial distance, initial velocity, and projection angle. Yet in the real world, balls, affected by air resistance, wind, and spin, do not fly in parabolas. Thus, the brain would further need to estimate, among other things, the speed and direction of the wind at each point of the ball's flight in order to compute the resulting path and the point where the ball will land. All this would have

Figure 1-1: How to catch a fly ball. Players rely on unconscious rules of thumb. When a ball comes in high, a player fixates his gaze on the ball, starts running, and adjusts his speed so that the angle of gaze remains constant.

to be completed within a few seconds—the time a ball is in the air. This is a standard account, according to which the mind solves a complex problem by a complex process. When it was tested experimentally, however, it turned out that players performed poorly in estimating where the ball would strike the ground.[6] If they were able to estimate, one would not see them running into walls, dugouts, and over the stands trying to chase fly balls. Clearly something else is at work.

Is there a simple rule of thumb that players use to catch balls? Experimental studies have shown that experienced players in fact use several rules of thumb. One of these is the *gaze heuristic*, which works in situations where a ball is already high up in the air:

Fix your gaze on the ball, start running, and adjust your running speed so that the angle of gaze remains constant.

The angle of gaze is the angle between the eye and the ball, relative to the ground. A player who uses this rule does not need to measure wind, air resistance, spin, or the other causal variables. All the relevant facts are contained in one variable: the angle of gaze. Note that a player using the gaze heuristic is not able to compute the point at which the ball will land. Yet the heuristic leads the player to the landing point.

As mentioned before, the gaze heuristic works in situations where the ball is already high in the air. If this is not yet the case, the player only needs to change the last of his strategy's three "building blocks":[7]

> Fix your gaze on the ball, start running, and adjust your running speed so that the image of the ball rises at a constant rate.

One can intuitively see its logic. If the player sees the ball rising from the point it was hit with accelerating speed, he'd better run backward, because the ball will hit the ground behind his present position. If, however, the ball rises with decreasing speed, he needs to run toward the ball instead. If the ball rises at a constant speed, the player is in the right position.

Now we can understand both how people catch fly balls without thinking and the cause of Phil's dilemma. Although the coach wrongly believed that players somehow calculate a trajectory, they in fact unconsciously rely on a simple rule of thumb that dictates the speed at which a player runs. Because Phil didn't understand why he did what he did either, he couldn't defend himself. Not knowing the rule of thumb can have unwanted consequences.

Most fielders are blithely unaware of the gaze heuristic, despite its simplicity.[8] Once the rationale underlying an intuitive feeling is made conscious, however, it can be taught. If you learn to fly an airplane, you will be instructed to use a version of this rule: When another plane approaches, and you fear collision, look at a scratch in your windshield and observe whether the other plane moves relative to that scratch. If it does not, dive away immediately. A good flight instructor will not ask anyone to calculate the trajectory of his plane in four-dimensional space (including time), estimate that of the other plane, and then see whether both trajectories intersect. Otherwise the pilot would likely not finish computing and realizing that a collision would take place before it in fact did. A simple rule is less prone to estimation or calculation error and is intuitively transparent.

The gaze heuristic and its relatives work for a class of problems that involve the interception of moving objects. In both ball games and pursuit, it helps to generate collisions, while in flying and sailing, it helps to avoid them.[9] Intercepting moving objects is an important adaptive task in human history, and we easily generalize the gaze heuristic from its evolutionary origins—such as hunting—to ball games. Interception techniques travel across species. From fish to bats, many organisms have the innate ability to track an object flying through three-dimensional space, a biological precondition for the gaze heuristic. The teleost fish catches its prey by keeping a constant angle between its line of motion and that of the target, and male hoverflies intercept females in the same way for mating.[10] When a dog goes after a sailing Frisbee, it is guided by the same instinct as the outfielder. In fact, a Frisbee has a more complicated flight path than a baseball; it curves in the air. By attaching a tiny camera to the head of a spaniel, a study showed that the dog ran so that the image of the ball was kept moving along a straight line.[11]

Interestingly, although the gaze heuristic works on an unconscious level, part of it appears in popular wisdom. For instance, when U.S. senator Russ Feingold noted that the Bush administration was clamping down on Iraq while Al-Qaeda was bubbling up else-where, he said: "I would ask you, Secretary Wolfowitz, are you sure we have our eye on the ball?"[12] Note that the gaze heuristic does not work for all interception problems. As many ballplayers say, the hardest ball to catch is the one that heads straight at you, a situation in which this rule of thumb is of no use.

The gaze heuristic exemplifies how a complex problem that no robot could match a human in solving—catching a ball in real time—can be easily mastered. It ignores all causal information rel-evant to computing the ball's trajectory and only attends to one piece of information, the angle of gaze. Its rationale is myopic, re-lying on incremental changes, rather than on the ideal of first computing the best solution and thereafter acting on it. Strategies relying on incremental changes also characterize how organiza-tions decide on their yearly budgets. At the Max Planck Institute where I work, my colleagues and I make slight adjustments to last year's budget, rather than calculating a new budget from scratch. Neither athletes nor business administrators need to know how to calculate the trajectory of the ball or the business. An intuitive "shortcut" will typically get them where they would like to be, and with a smaller chance of making grave errors.

DRUG COURIERS

Dan Horan always wanted to become a police officer, and even af-ter many years in this profession it has remained his dream job. His world is Los Angeles International Airport, where he tries to spot drug couriers. Couriers fly into LAX with hundreds of thou-sands of dollars in cash and fly out to other American cities, deliv-ering the drugs they've purchased. One summer evening in a terminal crowded with people waiting to board flights or meet in-coming passengers, Officer Horan circulated among them looking for something unusual. He was wearing shorts and a polo shirt, untucked to conceal his sidearm, handcuffs, and radio. To the untrained eye, there was nothing about him that suggested he was a police officer.

A woman arriving from New York's Kennedy Airport was neither untrained nor unwary.[13] She trailed a black rolling suitcase behind her, the color that nearly everyone today prefers. She had walked only twenty feet from the jetway door when her eyes met Horan's. In that instant, each formed an opinion about what business had brought the other to the airport, and both were right. Horan did not follow her beyond the escalator, but radioed his partner who was waiting outside the terminal. Horan and his partner were strikingly different in appearance. Horan is in his early forties and clean shaven, whereas his bearded partner was in his late fifties. Yet when the woman passed through the revolving doors into the baggage claim, it took her no longer than ten seconds to scan the crowd and recognize the partner for what he was. As the woman paced the terminal, a man who sat in a parked Ford Explorer just outside got out of the car and approached her. The woman spoke to him briefly, warning him of the detectives, then turned her back. The man returned to the car and drove off immediately, leaving her alone to face the police.

Horan's partner approached the woman, showed his police identification, and asked for her airline ticket. She did her best to conceal her unease, smiling and chatting, yet when the detective inquired about the contents of her suitcase, she appeared insulted and would not consent to a search of her luggage. She would have to accompany him to his office, the partner said, while he attempted to get a search warrant for her luggage. Over her fierce objections, she was handcuffed, and within a few minutes, a police dog had sniffed out the traces of drugs in her suitcase. A judge agreed to a search warrant, and when the officers opened the suitcase they found about two hundred thousand dollars in cash, which the woman admitted was intended for the purchase of a load of marijuana to be shipped to New York and sold on the street.

How did Horan intuitively pick this woman out of a crowd of several hundred? When I asked him, he didn't know. He could spot her in the large crowd, but could not spell out what seemed unusual about her. He was looking for someone who was looking for him. But what cues in her appearance made him believe that she was the courier? Horan could not say.

Although Horan's hunches allow him to excel in his job, the legal system does not necessarily approve. American courts tend to discount police officers' hunches, requiring them to articulate specific facts to justify a search, an interrogation, or an arrest. Even when an officer has a hunch, stops a car, finds illegal drugs or guns, and reports exactly this, judges often reject "mere hunches" as insufficient cause for a search.[14] They're trying to protect citizens from arbitrary searches and to protect their civil liberties generally. But their insistence on after-the-fact justification ignores that good expert judgment is generally of an intuitive nature. As a consequence, when police officers testify before a judge, they have learned not to use terms such as *hunch* or *gut instinct*, but to produce "objective" reasons after the fact. Otherwise, according to American law, all evidence discovered subsequent to a hunch might be suppressed, and the culprit might be acquitted.

Although many judges may condemn policemen's hunches, they tend to trust their own intuitions. As one judge explained to me, "I don't trust the police officers' hunches, because they are not my hunches." Similarly, prosecutors show little hesitation in justifying to themselves a peremptory challenge against a potential juror because she's wearing gold jewelry and a T-shirt or does not seem too bright, given that her hobbies are eating, doing her hair, and watching Oprah. However, the issue should be neither hunches per se nor the ability to come up with reasons after the fact while hiding the unconscious nature of hunches. To avoid discrimination, the legal system instead needs to survey the quality of policemen's hunches, that is, a detective's actual success in spotting criminals. In other professions, successful experts are evaluated by their performance rather than by their ability to give post-hoc explanations for their performance. Chicken sexers,[15] chess masters, professional baseball players, award-winning writers, and composers are typically unable to fully articulate how they do what they do. Many skills lack descriptive language.

UNCONSCIOUS INTELLIGENCE

Do gut feelings exist? The four preceding stories suggest that the answer is yes, and that both experts and laypeople rely on them. These stories are merely dots on the vast landscape of problems that intuition helps to solve: choosing partners, guessing trivia answers, catching fly balls, and detecting drug couriers. On many more occasions, intuition is the steering wheel through life. Intelligence is frequently at work without conscious thought. In fact, the cerebral cortex in which the flame of consciousness resides is packed with unconscious processes, as are the older parts of our brain. It would be erroneous to assume that intelligence is necessarily conscious and deliberate.[16] A native speaker can immediately tell whether a sentence is grammatically correct or not, but few can verbalize the underlying grammatical principles. We know more than we can tell.

Let me be clear about what a gut feeling is.[17] I use the terms *gut feeling, intuition,* or *hunch* interchangeably, to refer to a judgment

1. that appears quickly in consciousness,
2. whose underlying reasons we are not fully aware of, and
3. is strong enough to act upon.

But can we trust our guts? The answer to this question divides people into skeptical pessimists and passionate optimists. On the one hand, Sigmund Freud warned that it is "an illusion to expect anything from intuition," and many contemporary psychologists attack intuition as being systematically flawed because it ignores information, violates the laws of logic, and is the source of many human disasters.[18] In line with this negative spirit, our educational systems place value on everything but the art of intuition. On the other hand, ordinary people are inclined to trust their intuitions, and popular books eulogize the marvels of rapid cognition.[19] In this optimistic view, people generally know what to do, albeit not why. Optimists and pessimists tend to end up agreeing that hunches are often good, except when they are bad—which is true but not very helpful. Therefore the real question is not *if* but *when* can we trust our guts? To find the answer, we must figure out how intuition actually works in the first place.

What is the rationale underlying a gut feeling? Until recently, the answer to this question was largely unknown. By definition, the person with the feeling has no idea. Prominent philosophers have argued that intuition is mysterious and inexplicable. Can science lift the secret? Or does intuition elude human grasp—God's voice, lucky guesses, or a sixth sense beyond the limits of scientific understanding? In this book, I will argue that intuition is more than impulse and caprice; it has its own rationale. Let me first explain what I believe this rationale is *not*. When experiments such as the poster study demonstrated that in comparison to intuition, deliberate reasoning led to inferior outcomes, a big question mark arose: how can Franklin's balance sheet, the holy bible of decision theory, not work? Rather than challenging the sacred authority, researchers concluded that intuition must have performed the balance sheet method automatically, attending to all information and weighing it optimally, whereas conscious thinking had not done it properly.[20] Good choices must always be based on complex weighing of pros and cons, or so the conviction goes. But Franklin's moral algebra is not my vision of intuition, and, as we will see shortly, complexity is not always best.

How do I believe gut feelings work? Their rationale consists of two components:

1. simple rules of thumb, which take advantage of
2. evolved capacities of the brain.

I use the colloquial *rule of thumb* synonymously with the scientific term *heuristic*. A rule of thumb is quite different from a balance sheet with pros and cons; it tries to hit at the most important information and ignores the rest. For the million-dollar question, we know the rationale: the recognition heuristic, whose interesting feature is that it exploits one's partial ignorance. For catching a ball, we have identified the gaze heuristic, which ignores all information relevant to calculating a ball's trajectory. These rules of thumb enable fast action. Each takes advantage of an evolved capacity of the brain: recognition memory and the ability to track moving objects, respectively. The term *evolved* does not refer to a skill made by nature or nurture alone. Rather, nature gives humans a capability, and extended practice turns it into a capacity. Without evolved capacities, the simple rule could not do the job; without the rule, the capacities alone could not solve the problem either.

There are two ways to understand the nature of gut feelings. One is derived from logical principles and assumes that intuition solves a complex problem with a complex strategy. The other involves psychological principles, which bet on simplicity and take advantage of our evolved brain. Franklin's rule embodies the logical way: for each action, specify all consequences, weigh them carefully, and add the numbers up; then choose the one with the highest value or utility. Modern versions of this rule are known as maximizing the expected utility. This logical view assumes that minds function like calculating machines and ignores our evolved capacities, including cognitive abilities and social instincts. Yet these capacities come for free and enable fast and simple solutions for complex problems. The first goal of this book is to explicate the hidden rules of thumb underlying intuition, and the second is to understand when intuitions are likely to succeed—or fail. The intelligence of the unconscious is in knowing, without thinking, which rule is likely to work in which situation.

I have invited you on a voyage, but I must warn you: some of the insights we'll find on our trip conflict with the dogma of rational decision making. We will encounter doubt or outright disbelief at how accurate intuition can be and suspicion of its unconscious nature. Logic and related deliberate systems have monopolized the Western philosophy of the mind for too long. Yet logic is only one of many useful tools the mind can acquire. The mind, in my view, can be seen as an *adaptive toolbox* with genetically, culturally, and individually created and transmitted rules of thumb. Much of what I say is still controversial. Yet there is always hope. The U.S. biologist and geologist Louis Agassiz once said about new scientific insights: "First people say it conflicts with the Bible. Next, they say it has been discovered before. Lastly, they say they have always believed it." What I've written is based on my and others' research at the Max Planck Institute for Human Development and on that of many dear colleagues all around the world.[21] I hope that this little book will motivate readers to join us in exploring the new land of rationality.

Everything should be made as simple as possible, but not more so.

—Albert Einstein[1]

2 ⋮ LESS IS (SOMETIMES) MORE

The pediatric staff of a leading American teaching hospital is one of the best in the country. Years ago, the hospital admitted a twenty-one-month-old boy; let us call him Kevin.[2] Nearly everything was wrong with him: pale and withdrawn, he was drastically underweight for his age, refused to eat, and had constant ear infections. When Kevin was seven months old, his father moved out of the house, and his mother, who was often out "partying," sometimes missed feeding him altogether or tried to force-feed him jarred baby food and potato chips. A young doctor took charge of the case; he felt uncomfortable having to draw blood from this emaciated child and noticed that Kevin refused to eat after being poked with needles. Intuitively, he limited any invasive testing to the minimum, and instead tried to provide the child with a caring environment. The boy began to eat, and his condition improved.

The young doctor's superiors, however, did not encourage him in his unconventional efforts. Eventually, the young doctor could no longer impede the diagnostic machinery, and responsibility for Kevin was divided among a coterie of specialists, each interested in applying a particular diagnostic technology. According to their conception of medicine, the doctor's responsibility was to find the cause of the tiny boy's illness. They felt that they couldn't take chances: "If he dies without a diagnosis, then we have failed." Over the next nine weeks, Kevin was subjected to batteries of tests: CT scan, barium swallow, numerous biopsies and cultures of blood, six lumbar punctures, ultrasounds, and dozens of other clinical tests. What did the tests reveal? Nothing decisive. But under the bombardment of tests, Kevin stopped eating again. The specialists then countered the combined effects of infection, starvation, and testing with intravenous nutrition lines and blood infusions. Kevin died before his next scheduled test, a biopsy of the thymus gland. The physicians continued testing at the autopsy, hoping to find the hidden cause. One resident doctor commented after the boy died: "Why, at one time he had three IV drips going at once! He was spared no test to find out what was really going on. He died in spite of everything we did!"

THE BENEFITS OF FORGETTING

One day in the 1920s, the editor of a Russian newspaper met with his staff for their regular morning meeting. He read out the assignments for the day—lengthy lists of events and places to be covered, addresses, and instructions. While talking, he spotted a newly hired reporter who did not take notes. The editor was about to reproach him for not paying attention, when, to his surprise, the man repeated the entire assignment word for word. The reporter's name was Shereshevsky. Shortly after this event, the Russian psychologist A. R. Luria began to investigate Shereshevsky's fantastic memory. Luria read to him as many as thirty words, numbers, or letters at a time, and asked him to repeat these. Whereas ordinary humans can correctly repeat about seven (plus or minus two), the reporter recalled all thirty. Luria increased the elements to fifty, then to seventy, but the reporter recalled all correctly, and could even repeat them in reverse order. Luria studied him for three decades without being able to find any limits to his memory. Some fifteen years after their first meeting, Luria asked Shereshevsky to reproduce the series of words, numbers, or letters from that meeting. Shereshevsky paused, his eyes closed, and recalled the situation. They had been in Luria's apartment; Luria was wearing a gray suit and was sitting in a rocking chair and reading the series to him. Then, after all those years, Shereshevsky recited the series precisely. This was most extraordinary at the time, given that Shereshevsky had become a famous mnemonist who performed on stage and had been exposed to massive amounts of information to recall in each performance, which should have buried his old memories. Why did Mother Nature give perfect memory to him, and not to you and me?

There is a downside to such unlimited memory. Shereshevsky could recollect in detail virtually everything that had happened to him—both the important and the trivial. There was only one thing his brilliant memory failed to do. It could not forget. It was flooded by images of his childhood, for example, which could cause him acute malaise and chagrin. With a memory that was all details, he was unable to think on an abstract level. He complained of having a poor ability to recognize faces. "People's faces are constantly changing," he said, "it's the different shades of expression that confuse me and make it so hard to remember faces."[3] When reading a story, he could recite it word for word, but when asked to summarize the gist of the same story, he had to struggle. In general, when a task required going beyond the information given, such as understanding metaphors, poems, synonyms, and homonyms, Shereshevsky was more or less lost. Details that other people would forget occupied his mind and made it hard to move from this flow of images and sensations to some higher level of awareness about what was happei life—gist, abstraction, or meaning.

More memory is not generally better. Ever since Luria, prominent memory researchers have argued that the "sins" of our memory are necessary byproducts of a system adapted to the demands of our environments.[4] In this view, forgetting prevents the sheer mass of life's detail from critically slowing down the retrieval of relevant experience and so impairing the mind's ability to abstract, infer, and learn. Freud was an early advocate of adaptive forgetting. By repressing memories that include adverse emotional attributes or induce negative emotions when recalled, one can gain some immediate psychological advantage, he argued, even when the long-term costs of repression are harmful. The psychologist William James held a similar view when he said: "If we remembered everything, we should on most occasions be as ill off as if we remembered nothing."[5] A good memory is functional, and makes bets on what needs to be remembered next. The same functional principle is used in the file menu of many computer programs, such as Microsoft Word, where only the most recent items are listed. Word bets on the hypothesis that what users looked up last is likely what they will look up next (Figure 2-1).

Yet we needn't conclude that less memory is always better than perfect memory, or vice versa. The question is, what environmental structures make less than perfect memory desirable, and what structures favor perfect memory? I call this an ecological question, because it is about how cognition is adapted to its environment. What would a world look like in which a perfect memory is advantageous? One such world is that of the professional mnemonist into which Shereshevsky moved, where no abstraction is required. The philosophical world in which perfect memory would flourish is a completely predictable world, with no uncertainty.

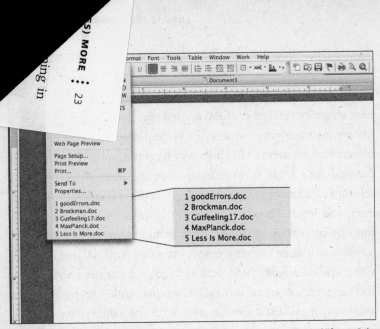

Figure 2-1: The Word program keeps only recently opened files in memory and "forgets" the rest. This usually speeds up finding what one is looking for.

THE IMPORTANCE OF STARTING SMALL

The world in which forgetting is adaptive is larger than we think. For people with painful and traumatic experiences, the ability to forget can produce relief. For children, the ability to forget seems to be essential for language learning. When Jeffrey Elman, a cognitive scientist, tried to get a large artificial neural network with extensive memory to learn the grammatical relationships in a set of several thousand sentences, the network faltered.[6] Instead of taking the obvious step of adding more memory to solve this problem, Elman restricted its memory, making the network forget after every three or four words—to mimic the memory restrictions of young children learning their first language. The network with the restricted memory could not possibly make sense of long, complicated sentences. But its restrictions forced it to focus on the short, simple sentences, which it did learn correctly, enabling it to master the small set of grammatical relationships in this subset. Elman then increased the network's effective memory to five or six words, and so on. By starting small, the network ultimately learned the entire corpus of sentences, which the full network with full memory had never been able to do alone. If parents read the *Wall Street Journal* to their babies and only talked to them using its sophisticated vocabulary, the babies' language development would probably be impeded. Mothers and fathers know this intuitively; they communicate with their infants in "baby talk" rather than by using elaborate grammatical structures. Limited memory can act like a filter, and parents unconsciously support this adaptive immaturity by providing limited input in the first place.

Starting small can be useful in areas beyond language development. For instance, a new business may develop more steadily with fewer people and less money than with a large group and a $10 million investment. Similarly, if a firm asks someone to do something spectacular, and pays him large sums of money, it may actually doom the project to failure. The rule "create scarcity and develop systematically" is a viable alternative in human as well as organizational development.

Cognitive limitations can help and hinder. It is easy to imagine situations where it is an advantage to start big. But cognitive limitations are not bad per se; they are only good or bad relative to the task at hand. The more complex a species, the longer the period of infancy tends to be. Humans are an extreme case, where a substantial fraction of their lifetime is spent in an immature state—physically, sexually, and mentally. One of our greatest minds, Albert Einstein, attributed his discovery of the theory of relativity to being a slow starter: "But my intellectual development was retarded, as a result of which I began to wonder about space and time only when I had already grown up. Naturally, I could go deeper into the problem than a child with normal abilities."[7]

WHEN ARE INVESTMENT INTUITIONS BETTER THAN OPTIMAL?

In 1990, Harry Markowitz received a Noble Prize in Economics for his pathbreaking work on optimal asset allocation. He addressed a vital investment problem that everyone faces in some form or another when saving for retirement or striving to earn money on the stock market. Say you are considering a number of investment funds. In order to reduce the risk, you don't want to put all your eggs in one basket. But how should you distribute the money over the various assets? Markowitz showed that there is an optimal portfolio that maximizes return and minimizes risk. When he made his own retirement investments, he surely relied on his Nobel Prize technique—or so one might think. But no, he didn't. He used a simple heuristic, the $1/N$ rule:

> Allocate your money equally to each of N funds.

Ordinary folks rely on the same rule intuitively—invest equally. In face, about half of the people in studies follow it; those who consider only two alternatives invest on a fifty-fifty basis, whereas most consider three or four funds and also distribute their money equally.[8] Isn't that intuition naive and financially foolish? Turning the question around, how much better is optimizing than $1/N$? A recent study compared a dozen optimal asset allocation policies, including that of Markowitz, with the $1/N$ rule in seven allocation problems.[9] The funds were mostly portfolios of stocks. One problem consisted of allocating one's money to the ten portfolios tracking the sectors comprising the Standard & Poor's 500 index, and another one to ten American industry portfolios. Not a single one of the optimal theories could outperform the simple $1/N$ rule, which typically made higher gains than the complex policies did.

To understand why less information and computation can be more, it is important to know that the complex policies base their estimates on existing data, such as the past performance of industry portfolios. The data fall into two categories, information that is useful to predicting the future, and arbitrary information or error that is not. Since the future is unknown, it is impossible to distinguish between these, and the complex strategies end up including arbitrary information. The formula $1/N$ would not be better than optimal policies in all possible worlds, however. These policies do best if they have data over a long time period. For instance, with fifty assets to allocate one's wealth, the complex policies would need a window of five hundred years to eventually outperform the $1/N$ rule. The simple rule, in contrast, ignores all previous information, which makes it immune to errors in the data. It bets on the wisdom of diversification by equal allocation.

CAN RECOGNITION OUTWIT FINANCIAL EXPERTS?

Does it pay to hire a renowned investment adviser for deciding which stocks to buy? Or is it better to save the consulting and managing fees and do it yourself, as long as you diversify? A forceful chorus of professional advisers warns that John Q. Public should not be left to his mere intuition: he cannot pick the stocks for himself, but needs their insider knowledge and sophisticated statistical computer programs to make money on the stock market. True?

In the year 2000, the investment magazine *Capital* announced a stock-picking contest. More than 10,000 participants, including the editor-in-chief, submitted portfolios. The editor laid down the rules: he chose fifty international Internet equities and set out a period of six weeks in which everyone could buy, hold, or sell any of these stocks in order to make profit. Many tried to gain as much information and insider knowledge about the stocks as possible, while others used high-speed computers to pick the right portfolio. But one portfolio stood out from all others.

This portfolio was based on collective ignorance rather than on expert knowledge and fancy software, and it was submitted by economist Andreas Ortmann and myself. We had looked for semi-ignorant people who knew so little about stocks that they had not even heard of many of them. We asked a hundred pedestrians in Berlin, fifty men and fifty women, which of the fifty stocks they recognized. Taking the ten stocks whose names were most often recognized, we created a portfolio. We submitted it to the contest in a buy-and-hold pattern; that is, we did not change the composition of the portfolio once it was purchased.

We hit a down market, which was not good news. Nevertheless, our portfolio based on collective recognition gained 2.5 percent. The benchmark proposed by *Capital* was its editor-in-chief, who knew more than all the hundred pedestrians together. His portfolio lost 18.5 percent. The recognition portfolio also had higher gains than 88 percent of all portfolios submitted, and beat various *Capital* indices. As a control, we had submitted a low-recognition portfolio with the ten stocks least recognized by the pedestrians, and it performed almost as badly as the editor-in-chief's. Results were similar in a second study, where we also analyzed gender differences. Interestingly, women recognized fewer stocks, yet the portfolio based on their recognition made more money than those based on men's recognition. This finding is consistent with earlier studies suggesting that women are less confident about their financial savvy, yet intuitively perform better.[10]

In these two studies, partial ignorance rather than extensive knowledge paid. Was that a one-off stroke of luck, as financial advisers are quick to suggest? As there is no foolproof investment strategy, name recognition will not always be a winner. We conducted a series of experiments that together suggest, however, that mere brand-name recognition matches financial experts, blue-chip mutual funds, and the market.[11] You may be asking whether I myself trust the collective wisdom enough to put my money where my mouth is. In one case I did and invested some fifty thousand dollars in a portfolio created by the name recognition of the most ignorant pedestrian group. After six months, the portfolio had gained 47 percent, better than the market and mutual funds managed by financial experts fared.

How can Jane and John Q. Public's collective ignorance be equal to renowned experts' knowledge? Peter Lynch, the legendary money manager of Fidelity's Magellan fund, gave exactly this advice to laypersons: invest in what you know. People tend to rely on the simple rule "Buy products whose brand name you recognize." This rule only helps if you are partially ignorant, that is, if you have heard of some and not all of the stocks. An expert such as *Capital*'s editor-in-chief who has heard of all the stocks cannot use it. In the United States alone, investment consultants earn about $100 billion annually in advising others how to play the market. Yet there is little evidence that advisers can predict much better than chance. On the contrary, some 70 percent of mutual funds perform below the market in any year, and none of the remaining 30 percent that happen to beat the market do so consistently.[12] Nevertheless, ordinary folks, companies, and governments pay the Wall Street clergy billions of dollars to tell them the answer to the big question "What is the market going to do?" As Warren Buffett, the billionaire financier, put it, the only value of stock forecasters is to make fortune-tellers look good.

THE ZERO-CHOICE DINNER

A few years ago, at Kansas State University, I gave a lecture on fast and frugal decision making. After a lively discussion, my kind host invited me out for dinner. He did not say where. The ride was long, too long, I thought. I guessed that he was taking me to a special restaurant, perhaps with a Michelin star or two. But in Kansas? Indeed, we were headed to a very special restaurant, albeit of a different kind. The Brookville Hotel was packed with people eager to dine, and when I sat down and looked at the menu, I knew why my host had taken me here. There was nothing to choose from. The menu listed exactly one item, the same one every day: half a skillet of fried chicken with mashed potatoes, cream-style corn, baking powder biscuits, and home-style ice cream. The people around me came from all over the place for the pleasure of not having to make a choice. And you can bet that the hotel knew how to prepare their only dinner; it was delicious!

The Brookville Hotel features a radical version of less-is-more—the zero-choice dinner. It embodies the reverse of the New York City ideal that more choice is always better, with menus that resemble encyclopedias rather than helpful guides. This idea that more choice is better flourishes beyond menus, fueling the lifeblood of much bureaucracy and commerce. At the beginning of the 1970s, Stanford University had two retirement plans that invested in stocks or in bonds. Around 1980, a third option was added, and a few years later, there were 5. By 2001, there were

Figure 2-2: Do customers buy more when there is more choice?

157 options.[13] Are 157 better than 5? Choice is good, and more choice is better, says the global business credo. According to rational choice theory, people weigh the costs and benefits of each alternative and pick the one they prefer. The more alternatives people have the higher the chance that the best one will be included and the more satisfied customers will be. But this is not how the human mind works. There is a limit to the information a human mind can digest, a limit that often corresponds to the magical number seven, plus or minus two, the capacity of short-term memory.[14]

If more choice is not always better, can it also hurt? Consider Draeger's Supermarket in Menlo Park, California, an upscale grocery shop known for its wide selection of foods. Draeger's features roughly seventy-five varieties of olive oil, two hundred and fifty varieties of mustard, and over three hundred varieties of jam. Psychologists set up a tasting booth inside the grocery store.[15] On the table were either six or twenty-four different jars of exotic jams. When did more customers stop? Sixty percent of the customers stopped for the wider selection, compared to 40 percent when fewer alternatives were offered. But when did more customers actually purchase any of the jams on offer? With twenty-four choices, only 3 percent of all shoppers bought one or more jars. However, when there were only six alternatives, 30 percent

...ht something. Thus, overall, ten times as many customers purchased the products when the selection was limited. Shoppers were more attracted by more alternatives, but many more bought the products when there was less choice.

Smaller selections can pay. Procter and Gamble reduced the number of versions of Head and Shoulders shampoo from twenty-six to fifteen, and sales increased by 10 percent. Unlike Draeger's, the global supermarket chain Aldi bets on simplicity: a small number of products bought in bulk, which allows for low prices, and an absolute minimum of service. Yet the quality of its products has a good reputation and is constantly monitored, much easier to accomplish with a small selection. Forbes estimated the fortune of the owners, the two Albrecht brothers, immediately behind that of Microsoft founder Bill Gates and the aforementioned Warren Buffett.[16] Is less choice also better for matters of the heart? An experiment with a group of young singles who were given online-dating profiles got the same pattern of results. These young people said that they would rather choose among twenty potential partners than among four. After having gone through this process, however, those given more choice found the situation less enjoyable and said that it did not increase satisfaction or reduce the feeling of having missed out on a better match.[17]

THE BEST POPS UP FIRST

A golfer has to go through a long list of steps when putting: judge the line of the ball, the grain of the turf, and the distance and angle to the hole; position the ball; align shoulders, hips, and feet to the left of the target; prepare for backswing; and so on. What advice should a coach give to a golfer? How about "Take your time, concentrate on what you are doing, and don't get distracted by anything around you." That seems like clever consulting to some,

and patently obvious to others; in any case, it is backed by research on the so-called *speed-accuracy trade-off*: the faster a task is
performed, the less accurate it becomes. And in fact, when novice
golfers are advised to take their time, concentrate on their movements, and focus their attention, they perform better. Should we
give expert golfers the same advice?

In an experiment, novices and expert golfers were studied under two conditions: they had either only up to three seconds for
each putt or all the time they wanted.[18] Under time pressure, as
mentioned, novices performed worse and had fewer target hits.
Yet surprisingly, experts hit the target more often when they had
less time than when they had no time limit. In a second experiment, players were either instructed to pay attention to their
swing or distracted by an unrelated, second task (counting tape-
recorded tones). When they were asked to pay attention to their
swing, as one might expect, novices did better than when they
were distracted. Yet with experts, it was again the opposite. When
experts concentrated on their swings, their performance decreased; when experts' attention was distracted, their performance
actually improved.

How can we account for this apparent paradox? Expert motor
skills are executed by unconscious parts of our brains, and conscious thinking about the sequence of behaviors interferes and becomes detrimental to performance. Setting a time limit is one
method to make thinking about the swing difficult; providing a distracting task is another. Since our conscious attention can focus on
only one thing at a time, it is fixed on the distracting task and cannot interfere with the swing.

Golf is not the only sport where taking time can hurt an expert. Indoor handball is a team sport in which players face a constant stream of quick decisions about what to do with the ball.
Pass, shoot, lob, or fake? Allocate the ball to the left wing player,

or to the right? Players have to make these decisions in an instant. Would they make better decisions if they had more time and could analyze the situation in depth? In an experiment with eighty-five young, skilled handball players, each stood in front of a screen, dressed in his uniform with a ball in his hand. On the screen, video scenes of high-level games were shown.[19] Each scene was ten seconds long, ending in a freeze-frame. The players were asked to imagine that they were the player with the ball, and when the scene was frozen, to name as quickly as possible the best action that came to mind. After their intuitive judgments, the players were given more time to inspect the frozen scene carefully, and asked to name as many additional options as they could. For instance, some discovered a player to the left or right they had overlooked, or noticed other details they were not aware of under time pressure. Finally, after forty-five seconds, they were asked to conclude what the best action would be. This final judgment was in about 40 percent of all cases different from the first choice. How did their intuitive first choices compare with the final decision upon reflection? To measure the quality of actions, professional-league coaches evaluated all proposed actions for each video. The hypothesis of a speed-accuracy trade-off suggests that when players have more time, they choose better actions because they have more information. As with the expert golfers, however, the opposite was the case. Taking time and analyzing did not generate better choices. In contrast, the gut reaction was, on average, better than the action chosen after reflection.

Why was the gut feeling so successful? Figure 2-3 shows the answer. The order in which possible actions came to players' minds directly mirrored their quality: the first action was substantially better than the second, which in turn was better than the third, and so on. Thus, having more time to generate options

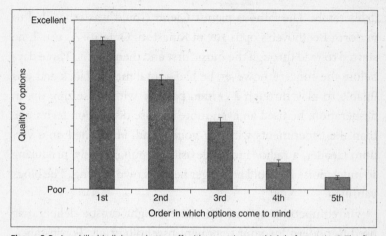

Figure 2-3: Are skilled ballplayers better off with more time to think before acting? The first spontaneous option that came to mind was the best one; the others were inferior (based on Johnson and Raab, 2003). Therefore, experienced players are well advised to follow their first gut feeling.

opens the door for inferior ones. This ability to generate the best options first is characteristic of an experienced player. Inexperienced players, by contrast, will not automatically generate the best actions first, and for them, more time and reflection may help. That the best option tends to pop up first has been reported for experts in various fields, such as firefighters and pilots.[20]

The speed-accuracy trade-off is one of psychology's well-established more-is-better principles. This earlier research was generally done with naive students rather than experts, however, and as we have seen, *more* (time, thought, attention) *is better* doesn't apply to expertly mastered skills. In such cases, thinking too much about the processes involved can slow down and disrupt performance (just think about how you tie your shoes). These processes run best outside of conscious awareness.

Stop thinking when you are skilled—this lesson can be applied deliberately. The famous pianist Glenn Gould was scheduled to perform Beethoven's opus 109 in Kingston, Ontario. As usual, he started to read through the music first and then play it. Three days before the concert, however, he had a total mental block and was unable to play through a certain passage without seizing up. In desperation, he used an even more intense distraction technique than the experiments with the golfers had. He turned on a vacuum cleaner, a radio, and a television simultaneously, producing so much noise he could no longer hear his own playing. The block vanished.

In competitive sports, the same insight can be deliberately used to undermine your opponent psychologically. For instance, while switching courts, ask your tennis opponent what he is doing to make his forehand so brilliant today. You have a good chance of making him think about his swing and weakening his forehand.[21] In sports, emergency units, and military actions, decisions need to be made fast, and striving for perfection by prolonged deliberation can lose the game or somebody's life. An ad for a computer game featuring U.S. covert operations in the Pacific theater in 1942 once caught my eye. It showed a picture of two marines on a road, facing a misty landscape with trees, bushes, and a wooden bridge across the road. Four locations were marked, and the question was asked: "What is the enemy hiding behind?" Having carefully inspected all the locations, I suddenly noticed the solution printed upside down beneath the picture: "You took too long to answer. You're dead."

MORE IS NOT ALWAYS BETTER

Gut feelings are based on surprisingly little information. That makes them look untrustworthy in the eyes of our superego, which has internalized the credo that more is always better. Yet experiments demonstrate the amazing fact that less time and information can improve decisions. *Less is more* means there is some range of information, time, or alternatives where a smaller amount is better. It does not mean that less is necessarily more over the total range. For instance, if one does not recognize any alternative, the recognition heuristic cannot be used. The same holds for choices between alternatives. If more people buy jam when there are six as opposed to twenty-four varieties, that does not imply that even more will buy when there are only one or two alternatives. Typically, there is some intermediate level where things work best. Less is more contradicts two core beliefs held in our culture:

More information is always better.
More choice is always better.

These beliefs exist in various forms and seem so self-evident that they are rarely stated explicitly.[22] Economists make an exception when information is not free: more information is always better unless the costs of acquiring further information surpass the expected gains. My point, however, is stronger. Even when information is free, situations exist where more information is detrimental. More memory is not always better. More time is not always better. More insider knowledge may help to explain yesterday's market by hindsight, but not to predict the market of tomorrow. Less is truly more under the following conditions:

A beneficial degree of ignorance. As illustrated by the recognition heuristic, the gut feeling can outperform a considerable amount of knowledge and information.

Unconscious motor skills. The gut feelings of trained experts are based on unconscious skills whose execution can be impeded by overdeliberation.

Cognitive limitations. Our brains seem to have built-in mechanisms, such as forgetting and starting small, that protect us from some of the dangers of possessing too much information. Without cognitive limitations, we would not function as intelligently as we do.

The freedom-of-choice paradox. The more options one has, the more possibilities for experiencing conflict arise, and the more difficult it becomes to compare the options. There is a point where more options, products, and choices hurt both seller and consumer.

The benefits of simplicity. In an uncertain world, simple rules of thumb can predict complex phenomena as well as or better than complex rules do.

Information costs. As in the case of the pediatric staff at the teaching hospital, extracting too much information can harm a patient. Similarly, at the workplace or in relationships, being overly curious can destroy trust.

Note that the first five items are genuine cases of less is more. Even if the layperson gained more information or the expert more time, or our memory retained all sensory information, or the company produced more varieties, all at no extra cost, they would still be worse off across the board. The last case is a trade-off in which it is the costs of further search that make less information the better choice. The little boy was hurt by the continuing diagnostic procedures, that is, by the physical and mental costs of search, not by the resulting information.

Good intuitions ignore information. Gut feelings spring from rules of thumb that extract only a few pieces of information from a complex environment, such as a recognized name or whether the angle of gaze is constant, and ignore the rest. How does this work, exactly? The next chapter provides a more detailed look at the mechanisms that allow us to focus on these few important pieces of information and ignore the rest.

It is a profoundly erroneous truism, repeated by all copy-books and by eminent people when they are making speeches, that we should cultivate the habit of thinking of what we are doing. The precise opposite is the case. Civilization advances by extending the number of important operations which we can perform without thinking about them.

—Alfred North Whitehead[1]

3 : HOW INTUITION WORKS

Charles Darwin thought of the hive bee's art of cell making as "the most wonderful of all known instincts."[2] He thought that this instinct evolved from numerous successive and slight modifications of simpler instincts. I believe that the evolution of cognition can be understood in a similar way, based on an adaptive toolbox of "instincts," which I call rules of thumb, or heuristics. Much of intuitive behavior, from perceiving to believing to deceiving, can be described in the form of these simple mechanisms that are adapted to the world we inhabit. They help us master the primary challenge for human intelligence: to go beyond the information given.[3] Let us begin with how our eyes and brains make unconscious bets.

BRAINS MAKE THINGS UP

King Henry VIII is known as having been a self-centered and forever mistrustful ruler who went through six marriages, with two of his wives joining the large tally of eminent persons he executed for alleged treason. As the story goes, his favorite dinner enjoyment was to close one eye and behead his guests. Would you like to give it a try? Close your right eye and stare at the smiling face located on the top right-hand side of Figure 3-1. Hold the book about ten inches away from your face, then move the book slowly toward you and away again, keeping your left eye focused on the smiling face. At some point, the sad face on the left will disappear, as if beheaded. Why does our brain act like a guillotine? The region in which the face disappears corresponds to the "blind spot" in the retina of the human eye. The eye acts like a camera, with a lens that directs light rays so that a picture of the world is created on the retina. The sheet of photoreceptors on the retina is much like a sheet of film at the back of the camera. But unlike film, there is a hole in it, through which the optical nerves exit the retina to transport the information

Figure 3-1: Seeing is betting. Close your right eye and stare at the smiling face in the top panel. Move the page closer to you while still staring; at some point the sad face on the left will disappear. Repeat the procedure with the bottom panel. At some point, your mind will repair the broken fork on the left. This creative process illustrates that the nature of perception is unconscious betting, not a veridical picture of what is out there.

to the brain. Because the hole has no photoreceptors, objects that would be processed in this region cannot be seen. When you look around with one eye closed, you might therefore expect to see a blank slate corresponding to this blind spot. In fact, you won't notice a thing. Our brains "fill in" the blank slate with a good guess. In Figure 3-1 (top), the best guess is "white" because the surrounding field is white. That guess makes the sad face disappear. In the same way, Henry VIII "beheaded" his guests by centering the image of their heads in his open eye's blind spot.

Now try something with your brain more constructive than beheading. Close your right eye and stare at the smiling face at the lower part of Figure 3-1, then move the book slowly toward you and away again. You will see that the broken fork on the left is miraculously repaired. The brain again makes its best guess based on the surrounding information: an elongated object crosses the blind spot from one side and continues on the other, thus it is likely that it exists in between. As in the case of the beheaded guest, these intelligent inferences are unconscious. Our brain cannot help but draw inferences about the world. Without them, we would see details but no structures.

Evolution could have created a better design in which the optical nerves exited from the backside rather than the surface of the retina. And in fact it has, but not for us. An octopus has no blind spot. The cells that carry information to the brain are located in the outer portions of the retina, so that the optic nerves do not have to cross through the retina. But even if evolution had favored us instead of the octopus, the general point remains, as the next section will illustrate. A good perceptual system has to go beyond the information given; it has to "invent" things. Your brain sees more than what your eye sees. Intelligence means making bets, taking risks.

I believe that intuitive judgments work in the same way as these perceptual bets. When given insufficient information, the brain makes things up based on assumptions about the world. The difference is that intuition is more flexible than perception. Let us first see how exactly these perceptual inferences work.

UNCONSCIOUS INFERENCES

To understand in more detail how our brains "go beyond the information given," consider the dots on the left-hand side of Figure 3-2. They appear concave; that is, they recede into the surface like small dents. The dots on the right-hand side, however, appear convex; that is, they project up from the surface, extending toward the observer. When you turn the book upside down, the concave dots will change into convex dots, and vice versa. Why do we see the dots this way or the other?

The answer is again that the eye does not have sufficient information to know for certain what is out there. But our brains are not paralyzed by uncertainty. The brain makes a "bet" based on the structure of the environment, or what the brain assumes the structure to be. Assuming a three-dimensional world, it uses the shaded parts of the dots to guess in what direction of the third dimension they extend. In order to make a good guess, it assumes that

 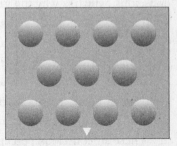

Figure 3-2: Unconscious inferences. The mind automatically infers that the dots in the left-hand picture are curved inward, that is, away from the observer, and those in the right-hand picture are curved outward, that is, toward the observer. If you turn the book around, the inward dots will pop out and vice versa.

1. light comes from above, and
2. there is only one source of light.

These two structures are characteristic in human (and mammalian) history, where sun and moon were the only sources of light. The first regularity also holds approximately true for artificial light today, which is typically placed above us—although exceptions, such as car lights, exist. The brain goes beyond the little information it has and relies on a simple rule of thumb adapted to these assumed structures:

> If the shade is in the upper part, then the dots recede into the surface; if the shade is in the lower part, then the dots project up from the surface.

Consider the dots on the right. They are bright in the upper part, and shaded in the lower part. Thus, the brain's unconscious inference is that the dots extend toward the observer, with light hitting the upper part and less light hitting the lower part. In contrast, the dots on the left side are shaded in the upper part and bright in the lower part; for the same reasons the brain bets that they must be curved inward. These assumptions, however, are generally not conscious, which is why the great German physiologist Hermann von Helmholtz spoke of *unconscious inferences*.[4] Unconscious inferences weave together data from the senses using prior knowledge about the world. There is a debate whether they are learned individually, as von Helmholtz and the Vienna psychologist Egon Brunswik argued, or are acquired by evolutionary learning, as the Stanford psychologist Roger Shepard and others have maintained.

These unconscious perceptual inferences are strong enough to act upon, but unlike other intuitive judgments, they are not flexible. They are triggered by external stimuli in an *automatic* way. An automatic process cannot be changed by insight or information external to the process. Even now, when we understand how the intuitive perception works, we cannot change what we see. We continue to see the concave dots suddenly pop out of the surface when we turn the book upside down.

Humans would not be called Homo sapiens if all inferences were like reflexes. As we have seen, other rules of thumb have all the advantages of perceptual bets—such as being fast, frugal, and adapted to their environment—but their use is not fully automatic. Although typically unconscious in nature, they can be subjected to conscious intervention. Consider how children infer the intentions of others.

WHAT DOES CHARLIE WANT?

From an early age, we intuitively have a feeling of what others want, what they desire, and what they think of us. But how do we arrive at these feelings? Let's show a child a schematic drawing of a face ("Charlie") surrounded by a seductive selection of chocolate bars (Figure 3-3).[5] We then say, "This is my friend Charlie. Charlie wants one of these treats. Which one does Charlie want?" How could a child possibly know? Yet almost all children immediately point to the same treat, the Milky Way. In contrast, children with autism tend to fail on this task. Some pick one, some another, and many egotistically pick their own favorite treat. Why do non-autistic children have a clear intuition about what Charlie wants, whereas autistic children do not? The answer is that the non-autistic children automatically engage in "mind reading." Mind readers can work with only minimal clues. They notice, but perhaps not consciously, that Charlie's eyes are pointed to the Milky Way, and so infer that this is the one he wants. However—and this is crucial—when asked what Charlie is looking at, children with autism answer correctly. What they do not seem to do as well

as other children is to make the spontaneous inference from looking to wanting:

> If a person looks at one alternative (longer than at others), it is likely the
> one the person desires.

In non-autistic children, this mind-reading heuristic is effortless
and automatic. It is part of their folk psychology. The ability to infer intentions from a gaze seems to be localized in the superior
temporal sulcus of our brain.[6] In children with autism this instinct
appears to be impaired. They don't seem to understand how the
minds of others work. In the words of Temple Grandin, an autistic woman who holds a PhD in animal science, much of the time
she feels "like an anthropologist on Mars."[7]

As with the unconscious inferences in perception, this simple
rule for inferring desire from gaze may well be anchored in our
genes and does not need much learning. However, unlike perceptual rules, the inference from gaze to desire is not automatic. If I
have reason to assume that Charlie wants to deceive me, I can
change my impression that he prefers the Milky Way. I might conclude that he is only looking at the Milky Way in order to influence

Figure 3-3: Which one does Charlie want?

me to take it, so that he can easily get the Snickers he in fact desires. Here we have a candidate for a rule of thumb that might be genetically coded and unconscious but nevertheless brought under voluntary control. In fact, autistic people sometimes use this voluntary control when trying to understand the secrets of mind reading. Grandin reported that, like a cognitive scientist, she tries to discover the rules that ordinary people use unconsciously and are unable to tell her. Then she uses a rule consciously, as if it were the grammar of a foreign language.

WHAT MAKES A GUT FEELING WORK?

Intuitive feelings seem mysterious and hard to explain—and most social scientists have steered clear of them. Even books that celebrate rapid judgments shy away from ever asking how a gut feeling arises. Rules of thumb provide the answer. They are typically unconscious but can be lifted to the conscious level. Most important, they are anchored both in the evolved brain and in the environment. By making use of both evolved capacities in our brain and environmental structures, rules of thumb and their product—gut feelings—can be highly successful. Let me walk you through this scheme.

- *Gut feelings* are what we experience. They appear quickly in consciousness, we do not fully understand why we have them, but we are prepared to act on them.
- *Rules of thumb* are responsible for producing gut feelings. For instance, the mind-reading heuristic tells us what others desire, the recognition heuristic produces a feeling of which product to trust, and the gaze heuristic generates an intuition of where to run.
- *Evolved capacities* are the construction material for rules of thumb. For example, the gaze heuristic takes advantage of the ability to

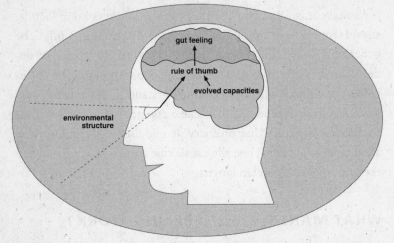

Figure 3-4: How gut feelings work. A gut feeling rapidly appears in consciousness, based on unconscious rules of thumb. These are anchored in the evolved capacities of the brain and the environment.

track objects. It is easy for humans—in contrast to robots—to track a moving object against a busy background; at three months old, babies have already begun to hold their gaze on moving targets.[8] Thus, the gaze heuristic is simple for humans but not for present-day robots.

- *Environmental structures* are the key to how well or poorly a rule of thumb works. For instance, the recognition heuristic takes advantage of situations where name recognition matches the quality of products or the size of cities. A gut feeling is not good or bad, rational or irrational per se. Its value is dependent on the context in which the rule of thumb is used.

Automatic rules, such as inferring depth from shade, and flexible rules, such as the recognition and gaze heuristics, all work according to this scheme. Yet there is an important difference. An automatic rule is adapted to our past environment without a present evaluation as to whether it is appropriate. It is simply triggered when a stimulus is present. Life has survived on this form of mindlessness from time immemorial. The flexible rules, in contrast, involve a quick *evaluation* of which one to use. If one doesn't work, there are others to choose from. The phrase "intelligence of the unconscious" refers to this quick process of evaluation. Brain imaging indicates that it might be associated with the anterior frontomedian cortex (see chapter 7). Gut feelings may appear simplistic, but their underlying intelligence lies in selecting the right rule of thumb for the right situation.

TWO WAYS TO UNDERSTAND BEHAVIOR

Like other approaches in the social sciences, the science of intuition attempts to explain and predict human behavior. Otherwise, it is unlike many of the other approaches. Gut feelings and rules of thumb are not the same kinds of explanations as fixed character traits, preferences, and attitudes, the key difference being that, as mentioned, rules of thumb are anchored not just in the brain but also in the environment. Explaining behavior in this way is what we call an adaptive approach, which assumes that people's behavior develops in a flexible way as they interact with their environment. Evolutionary psychology, for example, tries to understand present-day behavior by relating it to the past environment in which humans evolved.[9] Brunswik once compared the mind and the environment to a married couple who have to come to terms with one another. I will use his analogy to sketch out the difference between mental (internal) and adaptive explanations.

To generalize, let me distinguish between two ways a husband and wife can interact: to be kind and try to make each other happy, or to be nasty and try to hurt each other. Consider two couples, the Concords and the Frictions, who are in many respects similar. Yet the Concords are kind, warm, caring, and get along very well, while the Frictions fight, yell, insult each other, and are on the verge of splitting up. How can we explain the difference?

According to a widespread account, every person has a set of beliefs and desires, and these are the causes of behavior. For instance, Mr. and Ms. Friction might have sadomasochistic impulses and derive pleasure from hurting each other, and they simply maximize that pleasure. Alternatively, the couple might not have such desires but might instead have failed to calculate how they should behave. The first is the rational account and the second the irrational one, both assuming that people rely on mental calculations equivalent to Franklin's balance sheet. A third account is in terms of personality traits and attitudes, such as an overly aggressive temperament, or a dismissive attitude toward the other sex. Note that each of these explanations seeks the cause of a person's behavior in the individual mind. Personality theories investigate traits, attitude theories study attitudes, and cognitive theories focus on probabilities and utilities, or beliefs and desires.

The tendency to explain behavior internally without analyzing the environment is known as the "fundamental attribution error." Social psychologists have studied this tendency in the general public, but the same error creeps into social scientists' explanations as well. A person who takes financial risks with the stock market is not the same as one who takes social risks in dating or physical risks in mountain climbing. Few among us are risk seeking in all respects. As a student, I was fond of personality and attitude research but learned the hard way that it rarely predicts behavior well, and for good reason. The idea of fixed traits and preferences overlooks the adaptive nature of Homo sapiens. For the same reason, knowing the human genome does not mean understanding human behavior; the social environment also has a direct influence, possibly even on DNA's production of growth hormones.[10] As Brunswik observed, to understand the wife's behavior, one needs to find out what her husband is doing, and vice versa.

Adaptive theories focus on the relation between the mind and the environment, rather than on the mind alone.[11] Would that suggest a different story for the Concords and Frictions? Here we need to think about how rules of thumb interact with an environment's structure. What underlies the spouses' behavior? Consider a rule of thumb that has been dubbed *tit for tat*:

> Be kind first, keep a memory of size one, and imitate your partner's last behavior.

Assume that Ms. Concord, who unconsciously uses this rule of thumb, is solving a task with her husband for the very first time (caring for their first newborn, shopping for clothes together, or preparing dinner and doing the dishes). Ms. and Mr. Concord are kind to each other on the first occasion. The next time, she imitates his cooperative behavior, he imitates hers, and so on. The result can be a long harmonious relationship. The phrase "keep a memory of size one" means that only the last behavior (kind or nasty) is imitated and needs to be remembered. A relationship can grow if partners are willing to forget mistakes in the past, but not if one partner digs out the same old skeleton from the closet over and over again. In this case, forgetting means forgiving.

Most important, the same rule of thumb can lead to opposite behaviors, kind or nasty, depending on the social environment. If Ms. Concord had married someone with the maxim "Always be nasty to your wife, so that she knows who is the boss," her behavior would be the inverse. Triggered by her husband's nasty behavior, she would in turn react nastily to him. Behavior is not a mirror of a trait, but an adaptive reaction to one's environment.

Tit for tat works well if the partner also relies on it and does

not make errors. Assume that the Frictions also intuitively rely on
tit for tat. They similarly started out as a caring couple, but Mr.
Friction said something hostile in a burst of anger, and since then
there has been no end to exchanging blows. Ms. Friction was hurt,
so she responded with something similar. That made him hit back
at the next opportunity, and on it went. Now, the original incident
is long forgotten, but they have become trapped in an endless behavioral pattern. Mr. Friction feels that her last insult is to blame
for his own, and she thinks the same of her own response. How
can the Frictions stop the game, or not play it in the first place?
They could rely on a more forgiving rule called *tit for two tats*.

> Be kind first, then keep a memory of size two, and be nasty only if your
> partner did so twice; otherwise be kind.

Here, if he accidentally insults her, she gives him a second
chance. Only if it happens twice in a row does she retaliate. Tit for
two tats works better for couples in which one partner behaves
unreliably without being intentionally malicious. Yet its leniency
is vulnerable to exploitation. For instance, think of a man who gets
drunk and beats his wife one night, but deeply regrets it the following day and is gentle and thoughtful. If her mind works by tit
for two tats, she will remain kind to him. A shrewd man can consciously or unconsciously repeat the game for a long time to exploit
her sense of forgiving. Switching to tit for tat would mean that he
could no longer take advantage of her.

How well would these simple rules work in interactions with
many partners rather than one, who may rely on different kinds of
rules? In a widely publicized computer tournament, the American
political scientist Robert Axelrod had fifteen strategies compete
with each other, and the winnings were totaled up for all games.
The winning strategy was tit for tat, despite its simplicity.[12] In fact,

the most complex strategy was the least successful. Axelrod then worked out that if someone had submitted tit for two tats to his tournament, it would have been the winner instead. It is good at avoiding rounds of mutual recrimination, as happened with the Frictions. Does this mean that tit for two tats is generally better than tit for tat? No. Just as in real life, there is no single best strategy—it depends on the games the other players play. When Axelrod announced a second tournament, the eminent evolutionary biologist John Maynard Smith submitted tit for two tats. But the saintly heuristic did not win. Confronted with nasty strategies that tried to exploit the softies, it ranked far down the list. Once again the winner was tit for tat. Its wisdom is in its building blocks. In general, cooperation pays, forgetting pays, and imitation pays. And, most important, the combination pays. If one followed the biblical tenet "Turn the other cheek," exploitation would be a likely consequence.

Like the perceptual rules encountered in this chapter, tit for tat is based on evolved capacities, including those for imitation. These capacities are not the same as traits; rather, they are the stuff from which the rules of thumb are built. The next two chapters will address how they are anchored in the brain and in the environment, respectively.

> If we stopped doing everything for which we do not know the reason, or for which we cannot provide a justification . . . we would probably soon be dead.
>
> —Friedrich A. Hayek[1]

4 : EVOLVED BRAINS

We tend to smile when hearing that former first lady Barbara Bush is reported to have said, "I married the first man I ever kissed. When I tell this to my children, they just about throw up." Should she have looked at more suitors? Barbara Bush is not the only one; a third of Americans born even as recently as the 1960s and early 1970s married their first partner.[2] Marriage consultants often disapprove of people who marry the first or second partner they are engaged with, rather than looking systematically for more alternatives and experience in making such an important decision. Likewise, economists complain about the limited rationality in partner choice. When I hear similar criticisms, I ask the narrator how he found a partner. "Oh, that was different!" he tells me, and relates a story of an accidental meeting at a party or in a cafeteria, a first sensation of excitement, the anxiety of being rejected, how life focused on only that person, and a gut feeling that he or she was the right one. These stories have little in common with the deliberate process of choosing among a set of alternatives, as we tend to do when choosing digital cameras or refrigerators.

To date I have met only one man, an economist, who responded that he followed the Benjamin Franklin method to choose a partner. He sat down with a pencil and listed all the possible partners he could think of and all possible consequences he could imagine (such as whether she would still listen to him after being married, take care of the children, and let him work in peace). Next he put a number on the utilities of each consequence and then estimated the probabilities that each might come true. Finally, he multiplied the utilities with the probabilities and added them up. The woman he proposed to and married was the one with the highest expected utility, though he didn't tell her about his strategy. By the way, he is now divorced.

My point is that important decisions—whom to marry, which job to accept, what to do with the rest of your life—are not only a matter of our imagined pros and cons. Something else weighs in the decision process, something literally quite heavy: our evolved brain. It supplies us with capacities that have developed over millennia but are largely ignored by standard texts on decision making. It also supplies us with human culture, which evolves much faster than genes. These evolved capacities are indispensable for many important decisions and can prevent us from making crude errors in important affairs. They include the ability to trust, to imitate, and to experience emotions such as love. That is not to say that without trust and love, living beings would not function. In many reptiles, there is no mother love; newborn youngsters need to hide to prevent being eaten by their parents. That works too, but it's not how we humans behave. In order to understand human behavior, we need to understand that there is an evolved human brain that allows us to solve problems in our own way—different from that of reptiles or computer chips. Our infants don't need to hide after they are born but can draw on other abilities to grow up—smiling, imitating, looking cute, and having the capacity to listen and to learn to speak. Consider a thought experiment.

The Fable of Robot Love

In the year 2525, engineers finally managed to build robots that looked like humans, acted like humans, and were ready to reproduce. Ten thousand robots of various types had been built, all of them female. A research team set out to design a male robot who would be able to find a good mate, found a family, and take care of the little robots until they were able to take care of themselves. They called their first model Maximizer, M-1 for short. Programmed to find the best mate, M-1 proceeded to identify a thousand female robots that fit his goal of not marrying a model older than himself. He detected five hundred features on which individual female robots varied, such as energy consumption, computing speed, and frame elasticity. Regrettably, the females did not have their individual feature values printed on their foreheads; some even hid them, trying to fool M-1. He had to infer these values from samples of behavior. After three months passed, he had succeeded in getting reliable measures on the first feature he tested, memory size, from each female robot. The research team made a quick calculation of when M-1 would be ready to pick the best and discovered that no one in the team would still be alive at that point—nor would the best female robot. The thousand females were upset that M-1 could not make up his mind, and, as he began recording the second feature, the serial number, they pulled out his batteries and dumped him in a scrap yard. The team went back to the drawing board. M-2 was designed to focus on the most important features and to stop looking for more when the costs of collecting further information exceeded its benefits. After three months, M-2 was exactly where M-1 was, and in addition was busy measuring the benefits and costs of each feature so that he could know what to ignore. The impatient females ripped out his wires and got rid of him, too.

The team now adopted the proverb that the best is the enemy of the good and designed G-1, a robot who looked for a mate that was good enough. G-1 had an aspiration level built in. When he encountered the first female who met his aspiration level, he would propose to her—and ignore the rest. To make sure that he found a mate if his aspirations were too high, he was equipped with a feedback loop that lowered the aspiration level if none of the females were good enough for him over too long a period. G-1 showed no interest in the first six females he met, but then proposed to number seven. Short of alternatives, she accepted. Three months later, to everyone's pleasure, G-1 was married and had two small kids. While writing the final report, however, the team learned that G-1 had left his wife for another robot. Nothing in his brain had prevented him from running off to what looked to him like a better deal. One team member pointed out that M-1 would never have left his wife, because he would only have accepted the best in the first place. That's true, responded the others, but G-1 at least found one. The team discussed the problem for a while and then came up with GE-1. He was happy with a good-enough female, just like G-1, but was additionally equipped with an emotional glue that was released when he met a good-enough robot and adhered more strongly with each physical contact. Just to be sure, they inserted a second form of emotional glue into his brain that discharged when a baby was born and tightened after each physical contact with the baby. GE-1 proposed to a female as quickly as G-1 did, married, and fathered three babies. He was still with them when the team finished their report. He was somewhat clingy, but dependable. Ever since, GE-1 robots have conquered the earth.

In the fable, M-1 failed because he tried to find the best, as did M-2, both of them running out of time. G-1 was fast by going for a good-enough choice, but was also fast in dropping it. However, the capacity for love, the glue, provided a powerful stopping rule that ended GE-1's search for a partner and strengthened his commitment to his loved ones. Similarly, feelings of parental love, triggered by an infant's presence or smile, free parents from having to decide every morning whether they should invest their resources in their children or in some other business. The question of whether it is worthwhile to endure all the sleepless nights and other frustrations associated with baby care simply does not arise, and our memory ensures that we forget these tribulations soon. The evolved brain keeps us from looking too long and thinking too much. The culture it is embedded in influences what the object of love or trust can be, or what makes us upset and feel hurt.

Consider how a deliberate search for the best can conflict with pride and honor in real humans. The astronomer Johannes Kepler was short, unhealthy, and the son of a poor mercenary. Yet, famous for his astounding discoveries, he was considered a good catch. In 1611, after an arranged and unhappy first marriage, Kepler began a methodical search for a second wife. Unlike Barbara Bush, he investigated eleven possible replacements within two years. Friends urged him to marry candidate number four, a lady of high status and a tempting dowry, but he persisted with his investigation. Insulted, this suitable match rejected him for toying with her.

EVOLVED CAPACITIES

Evolved capacities, including language, recognition memory, object tracking, imitation, and emotions such as love, are acquired through natural selection, cultural transmission, or other mechanisms. The capacity for language, for instance, has evolved through natural selection, but knowing what words refer to what objects is a matter of cultural learning. I use the term *evolved capacities* in a broad sense, since capacities of the brain are always functions of both our genes and our learning environment. Historically, they evolved in tandem with the environment in which our ancestors lived and are shaped by the environment in which a child grows up. The human ability to imitate the behavior of others, for instance, is a precondition for the evolution of culture. One of Darwin's rare blunders was his belief that the ability to imitate was a common adaptation in animal species.[3] In no other species do individuals imitate as generally, carefully, and spontaneously as we do, allowing for the cumulative growth of a body of skills and knowledge that we call culture.

The psychologist Michael Tomasello and his coworkers conducted experiments wherein juvenile and adult chimpanzees and two-year-old children watched adult human demonstrators use a rakelike tool to obtain food that was out of arms' reach.[4] Chimps learned that the tool could somehow be used, but they didn't pay attention to the details of how it was used, whereas children paid close attention to the details and imitated them faithfully. A child may be weaker and slower than a chimp but learns culture faster using, in this case, imitation.

Yet if we all relied on imitation alone, our behavior would be decoupled from the environment. Flexible rules of thumb allow us to use imitation in an environmentally sensitive way. Imitate if the world is only slowly changing, otherwise learn from your own experience (or imitate those who are smarter than you and have adapted more quickly to a new situation).

Because many of our evolved capacities are not well understood, we can't endow machines with the same abilities. For instance, artificial face and voice recognition does not yet match that of humans, and emotional capacities, such as love, hope, and desire, are far from being part of machine intelligence. Of course, the opposite is also true. One can speak of the "evolved" capacities of modern computers, such as their immense combinatorial power unmatched by the human mind. The differences in the hardware of computer and that of mind have an important consequence. Humans and machines rely on different types of rules of thumb in order to take advantage of their respective capacities. Therefore, their intuitions are likely disparate.

Capacities build on each other. The ability to track objects is based on the physical and mental mechanisms involved in exploring one's environment. The ability to gather information by observing others is in turn based on the ability to track individuals across time and space. The capacities for cooperation and imitation are in turn based on the ability to observe others. If individuals have the ability to cooperate, in order to exchange goods for example, they'll also need to develop a radar for cheating in order to avoid being exploited.[5] Similarly, recognition memory is a precondition for reputation; for institutions to achieve good reputation, people must recognize their names and at least faintly recall why they deserve respect. An institution with a good reputation in turn increases trust, enhances group identification, and promotes the spread of the values it embodies.

THE ADAPTIVE TOOLBOX

Enlightenment philosophers compared the mind to a kingdom ruled by reason. At the turn of the twentieth century, William James compared consciousness to a river and the self to a fortress; and, in response to the latest technologies, the mind has been alternately portrayed as a telephone switchboard, a digital computer, and a neural network. The analogy I use is that of a toolbox that contains instruments adapted to the spectrum of problems confronting humankind (Figure 4-1). The adaptive toolbox has three layers:

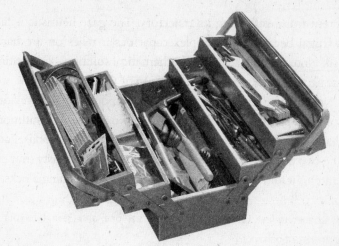

Figure 4-1: Like a maintenance worker with a box of tools, intuition draws on an adaptive toolbox of rules of thumb.

evolved capacities, building blocks that make use of capacities, and rules of thumb composed of building blocks. The relationship between these three layers can be compared to that between the atomic particles, the chemical elements in the periodic table, and the molecules built from combinations of the elements. There are many molecules and rules of thumb, fewer elements and building blocks, and even fewer particles and capacities.

Consider the gaze heuristic again. It has three building blocks:

(1) Fix your gaze on the ball, (2) start running, and (3) adjust your running speed so that the angle of gaze remains constant.

Each of these building blocks is anchored in evolved capacities. The first makes use of the human capacity to track objects, the second the capacity to maintain balance while running, and the third the capacity for fine-tuned vision-motor adjustments. These allow an original solution to the problem of catching a ball that is entirely different from estimating its trajectory. The gaze heuristic is fast and frugal because the complex capacities it relies on are hardwired. Note that the standard mathematical solution—computing the trajectory—does not take advantage of this potential.

Next consider the tit-for-tat strategy introduced in the previous chapter. It applies to situations in which two people or institutions exchange products, favors, emotional support, or other goods. Each of the two partners can be either nice (cooperate) or nasty (not cooperate). Tit for tat can also be divided into three building blocks:

> (1) Cooperate first, (2) keep a memory of size one, and (3) imitate your partner's last behavior.

Assume both partners meet repeatedly over time. Thus, in the first encounter, a person who uses tit for tat would be nice to the other, then remember how the other acted, and in the second encounter would imitate the partner's behavior in the first move, and so on. If the partner also uses tit for tat, both will cooperate from the beginning to the end; if the other is nasty and never cooperates, however, the tit-for-tat player will also finish by not cooperating with him. Note that although the resulting behavior is different, nice or nasty, the rule of thumb remains the same. An explanation of the tit-for-tat player's behavior in terms of traits or attitudes would miss this crucial difference between process (tit for tat) and resulting behavior (cooperate or not).

The first building block involves cooperation, the second involves the ability to forget, which, just like forgiving, is helpful for maintaining stable social relationships. My flash drive, by contrast, cannot forget, and from time to time I have to delete a number of files so that it remains useful. The third building block makes use of the ability to imitate, at which humans excel. Reciprocation between unrelated members of the same species is known as *reciprocal altruism*: I help you now, and you return the favor later. It is exceptionally rare in the animal world, as is tit for tat.[6] Animals may reciprocate if they are genetically related. By contrast, large human societies, which emerged only some ten thousand years ago, consist of mostly unrelated members who practice both nepotism and reciprocal altruism, as in agriculture and trade.

The adaptive toolbox consists of evolved capacities, including capacities to learn, that form the basis for building blocks that can construct efficient rules of thumb. Evolved capacities are the metal from which the tools are made. A gut feeling is like a drill, a simple instrument whose force lies in the quality of its material.

ADAPTIVE GOALS

An evolved capacity can be drawn upon to solve a wide spectrum of adaptive problems. Consider tracking. The original adaptive goals were probably predation and navigation: for example, intercepting prey by keeping the angle of gaze constant. As we have seen in chapter 1, tracking enables simple solutions to such complex modern problems as catching a baseball or avoiding a collision in sailing and flying. It provides ingenious solutions to social problems as well. In human societies as well as in hierarchically organized primate ones, a newcomer can quickly figure out the social status of individual group members by tracking who is looking at whom. Careful tracking allows a new group member to know whom to respect, avoiding conflicts that would upset the existing hierarchy. Children are sensitive to eye gaze from birth and seem to know when someone is looking at them. When infants are about one year old, they begin to use adults' gaze to learn language. When Mommy says *computer* while the child is looking at the goldfish tank, the child does not conclude that the new word refers to the tank or the fish but follows Mommy's gaze to infer which of the many objects in the room she means. At about the age of two, children begin to read others' gazes to figure out their mental states, such as their desires, and three-year-olds begin to use gaze as a cue to uncover deception.[7] Both children and adults trace not only the direction of a gaze, but also that of others' bodily movements to infer their intentions. Even the movements of virtual bugs on a computer screen can suggest to us that they might have intentions to flirt or help or hurt.[8]

An evolved capacity is necessary for solving adaptive problems, but it is not sufficient on its own—just as a 200-horsepower motor is designed for moving fast, but cannot do so without a steering wheel and tires. Only with those parts in place is it possible for a driver to move the car by a simple sequence of acts, such as turning on the engine, pushing the gas pedal, and switching gears accordingly. Similarly, the ability to track other people's eye gaze is not sufficient for inferring their intentions, as the case of autism illustrates. It is the rule of thumb that goes beyond the information given and forms our intuitions.

HUMAN AND MACHINE INTUITIONS

In 1945, the British mathematician Alan Turing (1912–54) predicted that computers would one day play excellent chess. Others have since hoped that chess programming would contribute to the understanding of how humans think. Though Turing was correct—in 1997, the IBM chess program Deep Blue beat the world champion Garry Kasparov—advances in programming have not led to a deeper understanding of human thought. Why is this? Human chess-playing strategies exploit the unique capacities of the human organism. Both Kasparov and Deep Blue had to rely on rules of thumb—even the fastest computer cannot determine the optimal strategy for chess, that is, the strategy that always wins, or at least never loses. Deep Blue can foresee as many as fourteen turns of play but has to use a quick rule of thumb in order to evaluate the quality of billions of possible positions generated. Kasparov, by contrast, is reported to have said that he thinks only four to five moves ahead. Deep Blue's capacities include its brute-force combinatorial power, whereas those of a grand master include spatial pattern recognition. Because these capacities are fundamentally different, understanding computer "thought processes" does not necessarily help to understand human ones.

In the initial wake of the computer revolution, the idea of disembodied cognition became very popular. Turing himself emphasized that differences in hardware were ultimately of little importance.[9] The new rhetoric was of cognitive systems that described the thought processes of "everything from man to mouse to microchip."[10] This led to great hopes for computer programs that would reproduce human creativity. A number of years ago, there was great enthusiasm about computer programs for composing music and improvising jazz, and anticipation that we could soon have a program that matched a Bach, if not a Beethoven. But nobody seriously talks anymore about simulating the great composers of the past. Unlike computer-generated music, human composition is embodied. It is based on an oral tradition of singing—wherein breath structures the phrasing and the length of tunes—and on the morphology of our hands, which structures the range and the flow of harmonies. And it is based on an emotional brain. Without the turbulent emotions Mozart faced as he wrote his *Ave Verum Corpus* on the eve of an early death, it is difficult to mimic his composition. Composition, like cognition, is based on capacities that vary from man to mouse to microchip.

HUMAN AND CHIMPANZEE INTUITIONS

Chimpanzee Intuitions

Humans are motivated, at least in part, by empathy and concern for the welfare of others. We donate blood for strangers, contribute to charity, and punish violators of social norms. Chimpanzees are, together with bonobos, our closest relatives, and they similarly engage in cooperative hunting, comfort victims of aggression, and perform other collective activities. Would they show concern for the welfare of unrelated, familiar chimps if the benefits were at no cost to themselves?

Primatologist Joan Silk and her collaborators conducted an experiment with chimps that had lived together for fifteen years or more.[11] Eighteen chimps were studied, from two different populations with different life histories and exposures to experiments. Pairs of chimps faced each other in opposing enclosures or sat side by side, and could see and hear each other. One chimp, the actor, was given the choice to pull one of two handles: if the actor pulled the "nice" handle, both the actor and the other chimp got food, and exactly the same portion. If the actor pulled the "nasty" handle, only the actor received food, and the other chimp got nothing. In a control test, only the actor was present. Which handle did the chimps pull?

When no other chimp was present, the actors chose both options about equally frequently. The chimps didn't care, and why should they? Yet even when a second chimp arrived, the chimps didn't choose the "nice" option more often. Although they could clearly see the other one displaying desperate begging gestures, or happily eating the food when it was dispensed, the chimps showed no sign of empathy. It should be noted that they showed no spitefulness either. What mattered to the actors more than the other chimp was whether the handle for the nice option was placed on their right or left side. They had a much stronger preference for the right side than for the happiness of their partner. Chimps simply did not seem to care about the welfare of unrelated group members.

Human Intuitions

What would children do in this situation? In a very similar study, three- to five-year-old children were asked whether they would prefer to have one sticker for themselves and one for a young female experimenter, or just a sticker for themselves.[12] Most children went with the prosocial alternative, and some were even willing to give up their own stickers to the experimenter.

In contrast to other primates, we humans not only give and share outside our families or when sharing proves costly, but we can get angry if someone does not. Consider the *ultimatum game*, which was invented by the economist Werner Güth, one of my colleagues at the Max Planck Society. In the game's classic version, two people who have never met before and never will are seated in different rooms. They cannot see or hear each other. A coin is flipped that assigns to them the role of the Proposer or the Responder. Both are told the rules of the game:

The Proposer receives $10 (in ten single bills) and offers any part of this to the Responder, that is, any amount from $0 to $10. The Responder then decides whether to accept the amount. If the Responder accepts, both players keep what they have; if the Responder rejects, both players receive nothing.

If you were the Proposer, how much would you offer? According to the logic of self-interest, both players will aim at maximizing their gain. Since the Proposer moves first, he should offer the Responder a single dollar bill and not more, because that maximizes the Proposer's gain. The Responder should subsequently accept the offer, because one dollar is obviously more than none. This logical norm is called a Nash equilibrium, after the Nobel laureate John Nash. But neither Proposer nor Responder tends to respond this way. The most frequent offer is not $1, but $5 or $4. Thus, people seem to be concerned with equity, sharing roughly the same amount—here we meet the 1/N rule in a different context. Even more surprising to the logic of self-interest, about half of those who were offered only one or two dollars rejected the money and preferred to take home nothing. They were annoyed and angry for being treated unfairly.

One might object that a few dollars are just peanuts, and that people would turn selfish as soon as there was more at stake. Imagine, for instance, a Proposer with $1,000 at his disposal. Yet when the game was played in other cultures with amounts that corresponded to the earnings of a week or even a month, little changed.[13] If the Proposer was a computer, humans were less likely to reject a small offer. Yet could the Proposer's concern for the other's welfare simply be calculated selfishness, that is, not wanting to take the risk of being rejected? Would people still give away money if the Responder could not reject? This version of the ultimatum game is called the dictator game, in which the Proposer simply dictates whether or not to give money and how much. Yet even when the other party has no possibility to reject, a substantial number of people give away some of their money. University students in the United States, Europe, and Japan playing the dictator game typically keep 80 percent and give 20 percent, whereas adults in the general population give more, sometimes an even split. German children's most frequent offer was an equal split in both games.[14] Nor could pure selfishness be found in a cross-cultural study with fifteen small-scale societies in the tropical forests of South America, the savanna-woodlands of Africa, the high-latitude deserts in Mongolia, and other remote places.[15] These experimental results illustrate that even in an extreme situation where the other person is unfamiliar, the situation anonymous, and there is a cost to themselves, people tend to be concerned with others' welfare. This general capacity for altruism divides us from other primates, even chimpanzees.

MALE AND FEMALE INTUITIONS

There is much talk about female intuition but comparatively little about that of males. One might suspect that this is because women have better intuitions than men, yet history suggests a different reason. Since the Enlightenment, intuition has been seen as inferior to reason, and long before that, women as inferior to men. Polarizing males and females in terms of both intelligence and character goes back to Aristotle, who wrote,

> The female is softer in disposition, is more mischievous, less simple, more impulsive, and more attentive to the nurture of the young; the male, on the other hand, is more spirited, more savage, more simple and less cunning. . . . The fact is, the nature of man is the most rounded off and complete, and consequently in man the qualities above referred to are found most clearly. Hence woman is more compassionate than man, more easily moved to tears, at the same time is more jealous, more querulous, more apt to scold and to strike. She is, fur-

thermore, more prone to despondency and less hopeful than the man, more void of shame, more false of speech, more deceptive, and of more retentive memory.[16]

This passage echoed through millennia of European debate about the difference between the genders and structured early modern views about moral values in Christianity. Violating the passive virtues, particularly chastity, was a cardinal sin for women but not for men, whereas timidity was easily excused in women but not in men. Memory, imagination, and sociability were traits clustered around the female pole, to be contrasted with male discursive and speculative reason. For Kant, this contrast condensed into the male mastery of abstract principles, as opposed to the female grasp of concrete detail, which in his view was incompatible with abstract speculations or knowledge: "Her philosophy is not to reason, but to sense."[17] He thought that the few flesh-and-blood counterexamples—learned ladies—were worse than useless, and freakish to boot: women with beards. A century later, Darwin similarly opposed male energy and genius to female compassion and powers of intuition. His identification of female faculties with "the lower races" was a characteristic nineteenth-century addition.

Modern psychology absorbed this opposition between male logic and female feeling into its initial concepts. Stanley Hall, the founder and first president of the American Psychological Association, described women as different from men in every organ and tissue:

She works by intuition and feeling; fear, anger, pity, love, and most of the emotions have a wider range and greater intensity. If she abandons her natural naiveté and takes up the burden of guiding and accounting for her life by consciousness, she is likely to lose more than she gains, according to the old saw that she who deliberates is lost.[18]

This short history reveals that the association between intuition and women has been, for much of the time, one between what were viewed as a lesser virtue and a lesser sex. Unlike the contrast between humans, chimpanzees, and machines, there is little firm evidence that men and women differ in any striking way in their cognitive capacities, except for characteristics associated with reproductive functioning and the cultures into which they are born. Given two millennia of beliefs in polar oppositions, however, it is not surprising that people believe that the differences between male and female intuition are larger than they really are. Psychologists tested the intuitive powers of more than fifteen thousand men and women in distinguishing a real smile from a false one.[19] They were shown ten pairs of photographs of smiling faces, one a genuine smile, the other a fake. Before studying the faces, the participants were asked to rate their intuitive abilities. Seventy-seven percent of the women said they were highly intuitive, compared to only 58 percent of the men. Yet women's intuitive judgments were not better than men's; they identified the real smile correctly in 71 percent of cases, whereas men did so in 72 percent. Interestingly, men could better judge women's genuine smiles than those of other men, whereas women were less adept at judging the sincerity of the opposite sex. Thus, if there are differences between male and female intuition, they are much more specific than the old idea that women are more intuitive than men.

For instance, according to the *selectivity hypothesis*, men tend to base their intuitive judgments on only one reason, good or bad, whereas women are sensitive to multiple reasons.[20] This difference has been attributed to societies in which girls are encouraged to consider others' views, whereas boys are motivated to take a more selfish, single-minded approach to mastering their world. Advertisers seem to assume this difference when they design ads for men and women. Researchers on consumer studies concluded that when targeting men, advertisers should associate the product with a single compelling message and feature it at the beginning of the ad. In contrast, when targeting women, they concluded that the ads should make use of ample cues that evoke positive associations and images. An automobile advertisement shows a Saab intently pursuing a straight path at a junction where large white arrows painted on the road point right and left. The headline reads: "Does popular acceptance require abandoning the very principles that got you where you are?" According to the ad, while other car manufacturers may compromise their design to win popular acceptance, Saab NEVER will! Never compromise is the single reason presented to buy a Saab. On the other hand, an ad by Clairol that introduced a new line of seven shampoos provided rich visual images meant to appeal to the female's powers of associative processing and subtle discriminations. Within a single ad, one shampoo was placed on a Hawaiian beach replete with luscious coconuts, another in a landscape of Egyptian pyramids near a desert oasis, and so on, for each of the seven products.

In 1910, seven years after she received the Nobel Prize in physics and a year before she became the first person ever to win a second Nobel Prize, this time in chemistry, Marie Curie was recommended for election to the prestigious French Academy of Sciences. In a tumultuous atmosphere, the members of the Academy voted and rejected her by a narrow margin. Despite her exceptional brilliance, prejudice against women prevailed; women, seen as inferior to men since the ancient world, were not meant to triumph in science. Today, though the polarity "male = reason, female = intuition" has been largely dissolved in our culture, and men are allowed to have intuitions, too, we still hear that women have much better intuition than men. Even now that intuition is seen as generally positive, this distinction sustains the old prejudice. Contrary to common belief, however, men and women share the same adaptive toolbox.

Human rational behavior is shaped by a scissors whose blades are the structure of task environments and the computational capabilities of the actor.

—Herbert A. Simon[1]

5 : **ADAPTED MINDS**

THE ANT ON THE BEACH

An ant rushes over a sandy beach on a path full of twists and turns. It turns right, left, back, then halts, and moves ahead again. How can we explain the complexity of the path it chose? We can think up a sophisticated program in the ant's brain that might explain its complex behavior, but we'll find that it does not work. What we have overlooked in our efforts to speculate about the ant's brain is the ant's environment. The structure of the wind-and-wave-molded beach, its little hills and valleys, and its obstacles shape the ant's path. The apparent complexity of the ant's behavior reflects the complexity of the ant's environment, rather than the ant's mind. The ant may be following a simple rule: get out of the sun and back to the nest as quickly as possible, without wasting energy by climbing obstacles such as sand mountains and sticks. Complex behavior does not imply complex mental strategies.

The Nobel laureate Herbert Simon argued that the same holds for humans: "A man, viewed as behaving systems, is quite simple. The apparent complexity of his behavior over time is largely a reflection of the complexity of the environment in which he finds himself."[2] In this view, people adapt to their environments much as gelatin does; if you wish to know what form it will have when it solidifies, also study the shape of the mold. The ant's path illustrates a general point: to understand behavior, one has to look at both the mind and its environment.

THE RAT IN THE MAZE

A lone, hungry rat runs through what psychologists call a T-maze (Figure 5-1, left). It can turn either left or right. If it turns left, it will find food in eight out of ten cases; if it turns right, there will only be food in two out of ten cases. The amount of food it finds is small, so it runs over and over again through the maze. Under a variety of experimental conditions, rats turn left most of the time, as one would expect. But sometimes they turn right, though this is the worse option, puzzling many a researcher. According to the logical principle called *maximizing*, the rat should always turn left, because there it can expect food 80 percent of the time. Sometimes, rats turn left in only about 80 percent of the cases, and right 20 percent of the time. Their behavior is then called *probability matching*, because it reflects the 80/20 percent probabilities. It results, however, in a smaller amount of food; the expectation is only 68 percent.[3] The rat's behavior seems irrational. Has evolution miswired the brain of this poor animal? Or are rats simply stupid?

We can understand the rat's behavior once we look into its natural environment rather than into its small brain. Under the natural conditions of foraging, a rat competes with many other rats and animals for food (the right-hand side of Figure 5-1). If all go to the spot that has the most food, each will get only a small share. The one mutant organism that sometimes chooses the second-best patch would face less competition, get more food,

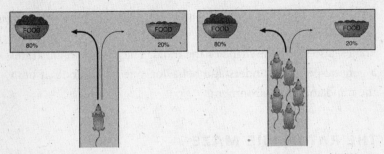

Figure 5-1: Rationality is in numbers. Rats run through a T-maze, in which they get food 80 percent of the time when they turn to the left and 20 percent of the time when they turn to the right. The single rat should turn left all the time, but it often does so only 80 percent of the time, even if this means getting less food. What looks like irrational behavior makes sense when there are many rats competing for limited resources; if all turned left, they would miss out on the food to the right.

and so be favored by natural selection. Thus, rats seem to rely on a strategy that works in a competitive environment but doesn't fit the experimental situation, in which an individual is kept in social isolation.

The stories of the ant and the rat make the same point. In order to understand behavior, one needs to look not only into the brain or mind but also into the structure of the physical and social environment.

CORPORATE CULTURE

New leaders galvanize companies with inspiring themes and ambitious plans, but they also influence corporate culture in simpler ways. Everyone has his or her personal rules of thumb, which they develop, often unconsciously, to help them make quick decisions. While leaders may not intentionally impose their own rules on the workplace, most employees implicitly follow them. These rules tend to become absorbed into the organizational bloodstream, where they may linger long after the leader has moved on. For example, if an executive makes it clear that excessive e-mail irritates her, employees unsure whether to include her in a message will simply opt against it. A leader who appears suspicious of employee absences discourages people from going to conferences or considering outside educational opportunities. Employees may be grateful that such shortcuts help them avoid protracted mulling over the pros and cons of taking a particular course of action. But as everyone adopts the same rules, the culture shifts: becoming more or less open, more or less inclusive, more or less formal. Because such behavior is difficult to change, leaders should think carefully about what values their rules communicate. They may even want to create new rules to shape the organization to their liking.

When I became director of the Max Planck Institute for Human Development, I wanted to create an interdisciplinary research group whose members actually talked, worked, and published together—a rare thing. Unless one actively creates an environment that supports this goal, collaboration tends to fall apart within a few years or may never get off the ground in the first place. The major obstacle is a mental one. Researchers, like most ordinary people, tend to identify with their ingroup and ignore or even look down on neighboring disciplines. Yet most relevant topics we study today do not respect the historically grown disciplinary borders, and to make progress one must look beyond one's own narrow point of view. So I came up with a set of rules—not verbalized, but acted upon—that would create the kind of culture I desired. Those rules included

Everyone on the same plane: In my experience, employees who work on different floors interact 50 percent less than those who work on the same floor, and the loss is greater for those working in different buildings. People often behave as if they still lived in the savanna, where they look for others horizontally but not above- or belowground. So when my growing group needed an additional two thousand square feet in which to operate, I vetoed the architect's proposal that we construct a new building and extended our existing offices horizontally, so that everyone remained on the same plane.

Start on equal footing: To ensure a level playing field at the beginning, I hired all the researchers at once and had them start simultaneously. That way no one knew more than anyone else about the new enterprise, and no one was patronized as a younger sibling.

Daily social gatherings: Informal interaction greases the wheels of formal collaboration. It helps to create trust and curiosity about what others do and know. To ensure a minimum daily requirement of chat, I created a custom. Every day at 4:00 p.m., everyone gathers for conversation and coffee prepared by someone from the group. Because there is no pressure to attend, almost everyone does.

Shared success: If a researcher (or a group) gets an award or publishes an article, he or she provides cake at coffee time. Note that the cake is not given to the successful person. He or she has to buy or bake it, turning everyone else into a beneficiary and sharing the success rather than creating a climate of envy.

Open doors: As director, I try to make myself available for anyone to discuss anything at any time. This open-door policy sets the example for other leaders, who make themselves equally accessible.[4]

All the members of the original group have since moved on to prestigious appointments elsewhere, but these rules have become an indelible part of who we are and a key to our successful collaboration. Many of the customs have assumed lives of their own—it has been years since I organized afternoon coffee times, yet somehow, every day, they still occur. I would advise all leaders to conduct a mental inventory of their own rules of thumb and to decide whether they want employees to be guided by them. The spirit of an organization is a mirror of the environment the leader creates.

THE STRUCTURE OF ENVIRONMENTS

The interplay between mind and environment can be expressed in a powerful analogy coined by Herbert Simon. In the epigraph to this chapter, mind and environment are compared to the blades of a pair of scissors. Just as one cannot understand how scissors cut by looking only at one blade, one will not understand human behavior by studying either cognition or the environment alone. This may seem to be common sense, yet much of psychology has gone a mentalist way, attempting to explain human behavior by attitudes, preferences, logic, or brain imaging and ignoring the structure of the environments in which people live.

Let us have a closer look at one important structure of the environmental blade: its uncertainty, that is, the degree to which surprising, novel, and unexpected things continue to happen. We cannot fully predict the future; usually, we are not even close.

UNCERTAINTY

Virtually every other morning I hear an interview on the radio in which a renowned financial expert is asked why certain stocks went up yesterday and others went down. The experts never fail to come up with a detailed, plausible account. The interviewer hardly ever asks the expert to predict which stocks will go up tomorrow. Hindsight is easy, foresight is hard. In hindsight, there is no uncertainty left; we know what has happened, and, if we are imaginative, we can always construct an explanation. In foresight, however, we must face uncertainty.

The stock market is an extreme example of an *uncertain environment*, with a predictability at or near chance level. The *Capital* stock-picking contest (chapter 2) that revealed the poor performance of financial experts was not a fluke. A recent study in Stockholm asked professional portfolio managers, analysts, brokers, and investment advisers to predict the performance of twenty blue-chip stocks. Two of the stocks were presented at a time, and the task was to predict which would perform better. A group of laypeople was given the same task, and their predictions were right 50 percent of the time. That is, as one might expect: laypeople performed at chance level, not better and not worse. How well did the professionals do? They picked the winning stock only 40 percent of the time. This result was replicated in a second study with another group of professionals.[5] How is it possible that financial experts' predictions were consistently worse than chance? The professionals base their predictions on complex information concerning each stock, and heavy competition leads them to create stock picks that vary widely from one to the next expert. Since not everyone can be right, this variability tends to decrease overall performance below chance.

Not all environments are as unpredictable as the stock market, but most are characterized by substantial unpredictability. Virtually nobody predicted the fall of the Berlin Wall, political scientists and the people of West and East Berlin included. Forecasters were surprised by the 1989 earthquake in California, the baby-boom population explosion, and the advent of the personal computer. Many don't seem to realize the limited predictability of our world, and both individuals and firms waste huge amounts of money on consultants. Each year, the "prediction industry"—the World Bank, stock brokerage firms, technology consultants, and business consulting firms, among others—earns some $200 billion as fortune-tellers, despite its generally poor track record. To predict the future is a challenge for laypeople, experts, and politicians alike. As Winston Churchill once complained, the future is one damn thing after another.[6]

SIMPLICITY AS AN ADAPTATION TO UNCERTAINTY

It is a common credo that in predicting the future, one should use as much information as possible and feed it into the most sophisticated computer. A complex problem demands a complex solution, so we are told. In fact, in unpredictable environments, the opposite is true.

High School Dropouts

Marty Brown is the father of two teenagers. He contemplates two high schools he might send his youngest son to, White High and Gray High. Haunted by his elder son's dropping out of school, Marty searches for a school with a low dropout rate. Neither school, however, provides reliable information about their dropout rates to the public. So Marty gathers information that could help him infer future dropout rates, including the schools' attendance rates, its writing scores, its social science test scores, the availability of English as a second language program, and class size. From his earlier experience with other schools, he has an idea of which of these are important clues. At some point, his intuition tells him that White High is the better choice. Marty feels strongly about his intuition and sends his youngest son to White High.

How likely is it that Marty's intuition is correct? In order to answer this question, we need to go through the scheme in Figure 3-4. First, we need to understand the rule of thumb that led to his intuitive feeling, and second, analyze in what environments this rule works. A number of psychological experiments suggest that people often, but not always, base their intuitive judgments on a single good reason.[7] A heuristic called *Take the Best* explains how a gut feeling can result from one-reason decision making. Let us assume that, like many experimental subjects, Marty relies on Take the Best. All he needs is a subjective feeling based on his experience with other schools as to which clues are better than others (this ranking need not be perfect). Assume that the top clues are attendance rate, writing score, and social science test score, in that order. The heuristic looks the clues up one by one and rates their values as *high* or *low*. If the first clue, the attendance rate, allows for a decision, then the process is stopped and all other information is ignored; if not, the second clue is checked, and so on. Here is a concrete illustration:

	Gray High or White High?	
Attendance rate?	high	high
Writing score?	low	high

STOP & PICK WHITE HIGH

The first clue, attendance rate, is not conclusive; thus the writing score is checked, and is conclusive. Search is stopped and the inference is made that White High has the lower dropout rate.

But how accurate is an intuition based on this rule of thumb? If Marty had used many more reasons, weighing and combining them as in Franklin's balance sheet method, would he have had a better chance of choosing the right school? I think it is fair to say that almost everyone believed the answer would definitely be yes before 1996, when my research group at the Max Planck Institute discovered the power of one-reason decision making.[8] Here is the short story.

Since the reasons for dropping out may vary between regions, let me concentrate on one large city: Chicago. To test the question of whether more clues are better than one good clue, we obtained information from fifty-seven schools on eighteen clues for dropout rates, including the proportion of low-income students, of students with limited English, of Hispanic students, and of black students; the average SAT score; the average income of the teachers; the parent participation rate; attendance rates; writing scores; social science test scores; and the availability of the English as a second language program. Now we were in a position to systematically study the problem Marty faced. How could we predict which of two schools has a higher dropout rate? According to Franklin's rule, we must consider all eighteen clues, weigh each carefully, and then make a prediction. The modern variant of Franklin's rule, made convenient by fast computers, is called multiple regression, where *multiple* stands for multiple clues. It determines the "optimal" weights for each clue, and adds them up, just like in Franklin's rule, but with complex computations. Our question was, how accurate is the simple Take the Best compared to this sophisticated strategy?

To answer this question, we did a computer simulation and fed in information on half of the schools—the eighteen clues and the schools' actual dropout rates. Based on this information, the

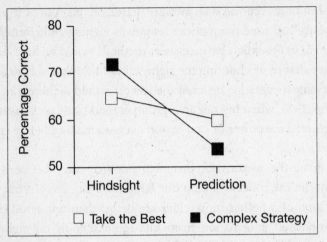

Figure 5-2: Intuitions based on a simple rule of thumb can be more accurate than complex calculations. How can we predict which Chicago high schools have higher dropout rates? If the facts for all high schools are already known (hindsight), the complex strategy ("multiple regression") does better; but if one has to predict dropout rates that are not yet known, the simple rule of thumb ("Take the Best") is more accurate.

complex strategy estimated the "optimal" weights and Take the Best estimated the order of clues. Then we tested both on the other half of the schools, using clues but no information about dropout rates.[9] This is called *prediction* in Figure 5-2 and corresponds to the situation Marty faces, in which he has experience with some schools but not with both of those he needs to choose from. As a control, we tested both strategies when all information about all schools was available. This *hindsight* task was to fit the data after the fact and therefore involved no prediction. What were the results?

The simple Take the Best predicted better than the complex strategy did (Figure 5-2), and it did so with less information. On average it looked at only three clues before it stopped, whereas the complex strategy weighed and added all eighteen. To explain what was already known about the schools (hindsight), the complex strategy was best. Yet when making predictions about what was not yet known, one good reason proved to be better than all reasons. If Marty's intuition follows Take the Best, he is more likely to make the right choice than if he carefully weighs and adds all of the available clues with a sophisticated computer program. This result teaches an important lesson:

> In an uncertain environment, good intuitions must ignore information.

But why did ignoring information pay in this case? High school dropout rates are highly unpredictable—in only 60 percent of the cases could the better strategy correctly predict which school had the higher rate. (Note that 50 percent would be chance.) Just as a financial adviser can produce a respectable explanation for yesterday's stock results, the complex strategy can weigh its many reasons so that the resulting equation fits well with what we already know. Yet, as Figure 5-2 clearly shows, in an uncertain world, a complex strategy can fail exactly because it explains too much in hindsight. Only part of the information is valuable for the future, and the art of intuition is to focus on that part and ignore the rest. A simple rule that relies only on the best clue has a good chance of hitting on that useful piece of information.

Relying on a simple strategy as opposed to a complex one not only has personal consequences for concerned parents like Marty; it can also affect public policy. According to the complex strategy, the best predictors for a high dropout rate were the school's percentage of Hispanic students, students with limited English, and black students—in that order. In contrast, Take the Best ranked attendance rate first, then writing score, then social science test score. On the basis of the complex analysis, a policy maker might recommend helping minorities to assimilate and supporting the English as a second language program. The simpler and better approach instead suggests that a policy maker should focus on getting students to attend class and teaching them the basics more thoroughly. Policy, not just accuracy, is at stake.

This analysis also provides a possible explanation for the conflict Harry experienced when trying to choose between his two girlfriends using the balance sheet (chapter 1). Mate choice, after all, involves a high degree of uncertainty. Perhaps Harry's feelings followed Take the Best, his heart going with the one most important reason. In that case, Harry's intuition may be superior to the complex calculation.

WHEN OPTIMIZATION IS OUT OF REACH

A solution to a given problem is called optimal if one can prove that no better solution exists. Some skeptics might ask, Why should intuition rely on a rule of thumb instead of the optimal strategy? To solve a problem by optimization—rather than by a rule of thumb—implies both that an optimal solution exists *and* that a strategy exists to find it. Computers would seem to be the ideal tool for finding the best solution to a problem. Yet paradoxically, the advent of high-speed computers has opened our eyes to the fact that the best strategy often cannot be found. Try to solve the following problem.

The Fifty-Cities Campaign Tour

A politician runs for president of the United States and plans to tour its fifty largest cities. Time is pressing, and the candidate travels in a convoy of automobiles. She wants to begin and end in the same city. What is the route with the shortest distance? The organizers have no idea. With a few more brains, couldn't they determine the shortest route? It seems an easy task. Simply determine all possible routes, measure the total distance, and choose the shortest. For instance, there are only twelve different routes for five cities.[10] Using a pocket calculator, it takes only minutes to determine the shortest route. However, with ten cities, there are already some 181,000 different routes, and the calculations become demanding. With fifty cities there are approximately

300,000,000,000,000,000,000,000,000,000,000,000,000,000,000,000,000,000,000,000,000,000

routes. Not even the fastest computer can check this many possibilities in a lifetime, in a century, or in a millennium. Such a problem is called "computationally intractable." In other words, we cannot determine the best route, however smart we are. Picking

Figure 5-3: The fifty-cities presidential campaign begins and ends in Boston. How can one find the shortest route? Good luck; not even the fastest computer can find the solution!

the best solution, optimization, is out of reach. What to do when optimization is no longer possible? Welcome to the world of rules of thumb. In this world, the question is, how does one find a good-enough solution? While we were pondering, the organizers had already finished planning the tour: the same as with the last candidate, with a few incremental changes due to closed highways.

Games

Consider tick-tack-toe. Player 1 makes a cross in one of a grid of nine squares, player 2 makes a circle in one of the empty squares, player 1 makes another cross, and so on. If a player manages to place three crosses or circles in a row (including diagonally), that person wins. In 1945, a robot was displayed in the entrance hall of the Museum of Science and Industry in Chicago, inviting the visitors to play a game of tick-tack-toe.[11] To their amazement, they

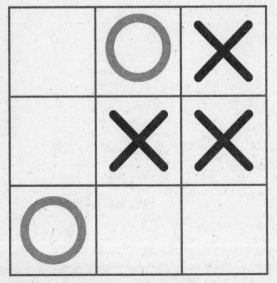

Figure 5-4: Tick-tack-toe. Can you find the best strategy?

never managed to beat the robot. It always won or tied because it knew the optimal solution to the game.

Player 1 makes a cross in the center square. If player 2 makes a circle in a middle square, as shown in Figure 5-4, player 1 makes a second cross in the adjacent corner square, which forces player 2 to sacrifice the next circle to prevent a row of three. Then player 1 makes a cross in the middle square adjacent to the other two crosses, which guarantees a win. Similarly, one can show that if the circle player makes the first circle in a corner, as opposed to a middle square, the cross player can always force a draw. This strategy either wins or draws, but never loses.

Here, the method to find the solution is enumeration and classification. For instance, for the first move, there are three options: center, corner, and middle square. All nine possibilities are in one of these three classes. The rest is enumeration of possibilities for the subsequent moves. By mere counting, one can prove that no other strategy does better. For simple situations such as tick-tack-toe, we know the best strategy. Good news? Yes and no: knowing the optimal strategy is exactly what makes the game boring.

Now consider chess. For each move, there are on average some thirty possibilities, which makes 30^{20} sequences in twenty moves, amounting to some

350,000,000,000,000,000,000,000,000,000

possible sequences of moves. This is a small number compared to that for the city tour problem. Can a chess computer determine the optimal sequence of moves for twenty moves? Deep Blue, the IBM chess computer, can examine some 200 million possible moves per second. At this breathtaking speed, Deep Blue would still need some fifty-five thousand billion years to think twenty moves ahead and pick the best one. (In comparison, the Big Bang is estimated to have occurred only some 14 billion years ago.) But twenty moves are not yet a complete game of chess. As a consequence, chess computers such as Deep Blue cannot find the best sequence of moves but have to rely on rules of thumb, just as grand masters do.

How do we know whether or not an optimal solution can be found for a game, or another well-defined problem? A problem is called "intractable" if the only known way of finding the perfect solution requires checking a number of steps that increase exponentially with the size of the problem. Chess is computationally intractable, as are the classic computer games Tetris, where one has to arrange a sequence of falling blocks, and Minesweeper.[12]

These games illustrate that even when problems are well defined, the perfect solution is often out of reach. A famous example from astronomy is the three-body problem. Three celestial bodies—such as earth, moon, and sun—move under no other influence than their mutual gravitation. How can we predict their movements? No general solution for this problem (or with four or more bodies) is known, whereas the two-body problem can be solved. With earthly bodies, even two defy a solution. There is no way to perfectly predict the dynamics of their mutual attraction, particularly if the attraction is emotional. In these situations, good rules of thumb become indispensable.

Ill-Defined Problems

Games such as chess and go have a well-defined structure. All permissible moves are defined by a few rules, nonpermissible moves can be easily detected, and what constitutes a victory is unequivocal. Victory in a political debate, on the other hand, is ill-defined. The set of permissible actions is not clearly defined, nor is what constitutes a winner. Is it better arguments, rhetoric, or one-liners? Unlike chess, a debate allows both candidates and their followers to claim victory. Similarly, in most bargaining situations (with the exception of auctions), the rules between buyers and sellers, employers and unions, are incompletely specified and need to be negotiated in the process. In everyday situations, rules are only partially known, can be overthrown by a powerful player, or are kept intentionally ambiguous. Uncertainty is prevalent; deception, lying, and lawbreaking possible. As a consequence, there are no optimal strategies known for winning a battle, leading an organization, bringing up children, or investing in the stock market. But, of course, good-enough strategies do exist.

In fact, people often prefer to retain a certain amount of ambiguity rather than to try to spell out all details. This is even true in legal contracts. The law in many countries assumes that a contract should spell out all possible consequences for the actions of each party, including punishment. Yet every smart lawyer knows that there is no perfectly watertight contract. Moreover, a large proportion of people who enter legal agreements feel that it is better to leave parts of the contract less well defined than they need be. As the legal expert Robert Scott argues, people may sense that there is no way to generate perfect certainty, but bet instead on the psychological factor of reciprocity, a powerful motivation for both sides in a contract. Trying to spell out all eventualities can seem like a lack of trust to the other partner and may actually do more harm than good.

The study of the match between the mind and its environment is still in its infancy. Prevailing explanations in the social sciences still look at only one of Simon's blades, focusing either on attitudes, traits, preferences, and other factors inside the mind or on external factors such as economic and legal structures. To understand what goes on inside our minds, we must look outside of them, and to understand what goes on outside, we should look inside.

We live in a dappled world, a world rich in different things, with different natures, behaving in different ways. The laws that describe this world are a patchwork, not a pyramid. They do not take after the simple, elegant and abstract structure of a system of axioms and theorems.

—Nancy Cartwright[1]

6 : WHY GOOD INTUITIONS SHOULDN'T BE LOGICAL

Imagine you are asked to participate in a psychological experiment. The experimenter gives you the following problem:

Linda is thirty-one years old, single, outspoken, and very bright. She majored in philosophy. As a student she was deeply concerned with issues of discrimination and social justice and participated in antinuclear demonstrations.

Which of the following two alternatives is more probable?

Linda is a bank teller
Linda is a bank teller and active in the feminist movement

Which one did you choose? If your intuitions work like those of most people, you picked the second alternative. Amos Tversky and Nobel laureate Daniel Kahneman, however, argued that this is the false answer, because it violates logic. A conjunction of two events (Linda is a bank teller *and* active in the feminist movement) cannot be more probable than only one of them (Linda is a bank teller). In other words, a subset can never be larger than the set itself. "Like it or not, A cannot be less probable than (A&B), and a belief to the contrary is fallacious."[2] They labeled the intuition shared by most people the *conjunction fallacy*. The Linda problem has been used to argue that human beings are fundamentally illogical and has been invoked to explain various economic and human disasters, including U.S. security policy, John Q. Public's fear of nuclear reactor failures, and his imprudent spending on insurance. The evolutionary biologist Stephen Jay Gould wrote,

> I am particularly fond of [the Linda] example, because I know that the [conjunction] is least probable, yet a little homunculus in my head continues to jump up and down, shouting at me— "but she can't just be a bank teller: read the description." . . . Why do we consistently make this simple logical error? Tversky and Kahneman argue, correctly I think, that our minds are not built (for whatever reason) to work by the rules of probability.[3]

Gould should have trusted the gut feeling of his homunculus, rather than his conscious reflections. Academics who agree with the conjunction fallacy believe that mathematical logic is the basis for determining whether judgments are rational or irrational. In the Linda problem, all that counts for the logical definition of rational reasoning are the English terms *and* and *probable*, which are assumed to have only one correct meaning: the logical AND (that we use, for example, in search machines) and mathematical probability (a comparison of the number of favorable outcomes to the number of possible outcomes). I call such logical norms *content-blind* because they ignore the content and the goals of thinking. Rigid logical norms overlook that intelligence has to operate in an uncertain world, not in the artificial certainty of a logical system, and needs to go beyond the information given. One major source of uncertainty in the Linda problem is the meaning of the terms *probable* and *and*. Each of these terms has several meanings, as any good English dictionary or its equivalent in other languages will reveal. Consider the meanings of *probable*. A few, such as "what happens frequently," correspond to mathematical probability, but most do not, including "what is plausible," "what is believable," and "whether there is evidence." As we have seen in chapter 3, perception solves this problem of ambiguity by using intelligent rules of thumb, and, I argue, so does higher-order cognition. One of these unconscious rules that our minds appear to use to understand the meaning of language is the conversational *maxim of relevance*.[4]

Assume that the speaker follows the principle "Be relevant."

The unconscious inference is thus: if the experimenter reads to me the description of Linda, it is most likely relevant for what he expects me to do. Yet the description would be totally irrelevant if one understood the term *probable* as mathematical probability. Therefore, the relevance rule suggests that *probable* must mean something that makes the description relevant, such as whether it is plausible. Read the description—Gould's homunculus understood this point.

Is most people's answer to the Linda problem based on a reasoning fallacy or on an intelligent conversational intuition? To decide between these alternatives, Ralph Hertwig and I asked people to paraphrase the Linda task for a person who is not a native speaker and does not know the meaning of *probable*. Most people used nonmathematical meanings such as whether it is possible, conceivable, plausible, reasonable, and typical. Only very few used "frequent" or other mathematical meanings. This suggests that conversational intuition rather than logical error is at issue, specifically the ability to infer the meaning of ambiguous statements by means of conversational rules. As a further test of this hypothesis, we changed the ambiguous phrase *probable* into a clear *how many?*

> There are a hundred persons who fit the description above (i.e., Linda's). How many of them are
>
> bank tellers?
> bank tellers and active in the feminist movement?

If people don't understand that a set cannot be smaller than a subset, and consistently make this logical blunder, then this new version should produce the same results as the old one. If on the other hand, there is no blunder but people make intelligent unconscious inferences about what meanings of *probable* make the description of Linda relevant, these meanings are now excluded and the so-called fallacy should largely disappear. And this indeed is what happened (Figure 6-1).[5] The result is consistent with earlier research by the Swiss psychologists Bärbel Inhelder and Jean Piaget, who performed similar experiments with children ("Are there more flowers or more primulas?") and reported that by the age of eight, a majority gave responses consistent with class inclusion. Note that children were asked how many, not how probable. It would be very strange if later in life adults could no longer understand what eight-year-olds can. Logic is not a sensible norm for understanding the question "Which alternative is more probable?" in the Linda problem. Human intuition is much richer and can make reasonable guesses under uncertainty.

The Linda problem—and the hundreds of studies it has generated to find out what conditions make people reason more or less logically—illustrates how the fascination with logic leads researchers to pose the wrong questions and miss the interesting, psychological ones. The question is not whether people's intuitions

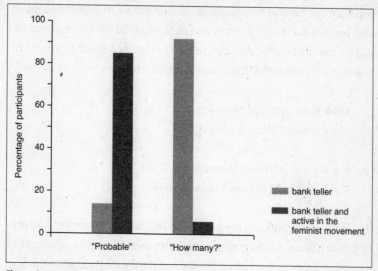

Figure 6-1: Is Linda a bank teller?

follow the laws of logic, but rather, what unconscious rules of thumb underlie intuitions about meaning. Let us take a closer look at natural language comprehension.

Peggy and Paul

In first-order logic, the particle AND is *commutative*, that is, *a AND b* is equivalent to *b AND a*. Yet again, this is not how we understand natural language. For instance, consider the following two sentences:

> Peggy and Paul married *and* Peggy became pregnant.
> Peggy became pregnant *and* Peggy and Paul married.

We know intuitively that the two sentences convey different messages. The first suggests that pregnancy followed marriage, whereas the second implies that pregnancy came first and was possibly the reason for marriage. If our intuition worked logically and treated the English term *and* as the logical AND, we wouldn't notice the difference. *And* can refer to a chronological or causal relation, neither of which is commutative. Here are two more pairs:

Mark got angry *and* Mary left.
Mary left *and* Mark got angry.

Verona is in Italy *and* Valencia is in Spain.
Valencia is in Spain *and* Verona is in Italy.

We understand in a blink that the first pair of sentences conveys opposite causal messages, whereas the second pair is identical in meaning. Only in the last pair is the *and* used in the sense of the logical AND. Even more surprising, we also know without thinking when *and* should be interpreted as the logical OR, as in the sentence

We invited friends *and* colleagues.

This sentence refers to the joint set of friends and colleagues, not to their intersection. Not everyone is both a friend and a colleague; many are either or. Once again, intuitive understanding violates the conjunction rule, but this is not an error of judgment. Rather it is an indication that natural language is more sophisticated than logic.

How do our minds infer at one glance what *and* means in each context? These inferences have the three characteristics of intuitions: I know the meaning, I act on it, but I do not know how I know it. Since a single sentence is sufficient as context, the clues must come from the content of the sentence. To this day, linguists are still working on spelling out the rules of thumb that underlie this remarkably intelligent intuition. No computer program can decode the meaning of an *and* sentence as well as we can. These are the interesting unconscious processes that we only partly understand, but which our intuition masters in the blink of an eye.

Framing

Framing is defined as the expression of logically equivalent information (whether numerical or verbal) in different ways. For example, your mother has to decide whether she will have a difficult operation and is struggling with the decision. Her physician says that she has a 10 percent chance of dying from the operation. That same day another patient asks about the same operation. He is told that he has a 90 percent chance of surviving.

Logic does not make a difference between either of these statements, and consequently, logically minded psychologists have argued that human intuition should be indifferent, too. They claim that one should ignore whether one's doctor describes the outcome of a possible operation as a 90 percent chance of survival (a positive frame) or a 10 percent chance of dying (a negative frame). But patients pay attention and try to read between the lines. By using a positive frame, the doctor might signal to the patient that the operation is the best choice. In fact, patients accept the treatment more often if doctors choose a positive frame.[6] Kahneman and Tversky, however, interpret attention to framing to mean that people are incapable of retranslating the two versions of the doctors' answer into a common abstract form and are convinced that "in their stubborn appeal, framing effects resemble perceptual illusions more than computational errors."[7]

I disagree. Framing can communicate information that is overlooked by mere logic. Consider the most famous of all framing examples:

The glass is half full.
The glass is half empty.

According to the logical norm, people's choices should not be affected by the two formulations. Is the description in fact irrelevant? In an experiment, a full glass of water and an empty glass are put on a table.[8] The experimenter asks the participant to pour half of the water into the other glass, and to place the half-empty glass at the edge of the table. Which one does the participant pick? Most people chose the previously full glass. When other participants were asked to move the half-full glass, most of them chose the previously empty one. This experiment reveals that the framing of a request helps people extract surplus information concerning the dynamics or history of the situation and helps them to guess what it means. Once again, intuition is richer than logic. Of course, one can mislead people by framing a choice accordingly. But that possibility does not mean that attending to framing is irrational. Any communication tool, from language to percentages, can be exploited.

The potential of framing is now being recognized in many disciplines. The renowned physicist Richard Feynman emphasized the importance of deriving different formulations for the same physical law, even if they are mathematically equivalent. "Psychologically they are different because they are completely unequivalent when you are trying to guess new laws."[9] Playing with different representations of the same information helped Feynman to make new discoveries, and his famous diagrams embody the emphasis he placed on presentation. Yet psychologists themselves are in danger of discarding psychology for mere logic.

THE CHAIN STORE PARADOX

Reinhard Selten is a Nobel laureate in economics who made the "chain store problem" famous by proving that an aggressive policy against competitors is futile. Here is the problem:

A chain store called Paradise has branches in twenty cities. A competitor, Nirvana, plans to open a similar chain of stores and decide one by one whether or not to enter the market in each of these cities. Whenever a local challenger enters the market, Paradise can respond either with aggressive predatory pricing, which causes both sides to lose money, or with cooperative pricing, which will result in sharing profits 50-50 with the challenger. How should Paradise react when the first Nirvana store enters the market? Aggression or cooperation?

One might think that Paradise should react aggressively to early challengers, in order to deter others from entering the market. Yet using a logical argument, Selten proved that the best answer is cooperation. His argument is known as *backward induction*, because one argues backward from the end to the beginning. When the twentieth challenger enters the market, there is no reason for aggression, because there is no future competitor to deter and so no reason to sacrifice money. Given that the chain store will decide to be cooperative to the last competitor, there is no reason to be aggressive to the nineteenth competitor either, because everyone knows that the final challenger cannot be deterred. Thus, it is rational for Paradise to cooperate with the next-to-last challenger, too. The same argument applies to the eighteenth challenger, and so on, back to the first. Selten's proof by backward induction implies that the chain store should always respond cooperatively, in every city, from the first to the last challenger.

But that is not the end of the story. Seeing his result, Selten found his logically correct proof intuitively unconvincing and indicated that he would rather follow his gut feeling to be aggressive in order to deter others from entering the market.

I would be very surprised if it failed to work. From my discussion with friends and colleagues, I get the impression that most people share this inclination. In fact, up to now I met nobody who said that he would behave according to [backward] induction theory. My experience suggests that mathematically trained persons recognize the logical validity of the induction argument, but they refuse to accept it as a guide to practical behavior.[10]

Those who don't know Reinhard Selten might suspect that he is an overly aggressive person whose impulses overwhelm his thinking, but he is not. The clash between Selten's logic and his intuitions have nothing to do with a preference for aggressive action. As we have seen repeatedly, logical arguments may conflict with intuition. And as we have also seen, intuition is often the better guide in the real world.

DISEMBODIED INTELLIGENCE

Artificial intelligence (AI) has spent some of its history building disembodied intelligences that carry out abstract activities, such as chess playing, where the only interaction with the world is by means of a screen or printer. This suggests that the nature of thinking is logical, not psychological. Logic is the ideal of a disembodied system, the proper yardstick for deductive arguments that concern the truth of propositions, such as in a mathematical proof. Few logicians, however, would argue that logic provides the yardstick for all kinds of thinking. Similarly, a century ago Wilhelm Wundt, who has been called the father of experimental psychology, pointed out the difference between the laws of logic and the process of thinking.[11]

At first it was thought that the surest way would be to take as a foundation for the psychological analysis of the thought-processes the laws of logical thinking, as they had been laid down from the time of Aristotle by the science of logic. These norms only apply to a small part of the thought-processes. Any attempt to explain, out of these norms, thought in the psychological sense of the word can only lead to an entanglement of the real facts in a net of logical reflections. We can in fact say of such attempts, that measured by the results they have been absolutely fruitless. They have disregarded the psychological processes themselves.

I could not agree more with Wundt. Nevertheless, as we have seen, many psychologists have treated forms of logic as a universal calculus of cognition, and many economists use it as a universal calculus of rational action. The work of Piaget, for instance, is concerned with the growth of all knowledge, from the intellectual growth of one child to human intellectual history. For him, the development of cognition was fundamentally the development of logical structures, from prelogical thought to abstract formal reasoning.[12] The ideal of logic is so embedded in our culture that even those who criticize Piaget's claim as empirically wrong often maintain it as the universal standard of good reasoning. Those who violate that standard are diagnosed as having cognitive illusions, such as the conjunction fallacy.

Generations of students in the social sciences have been exposed to entertaining lectures that point out how dumb everyone else is, constantly wandering off the path of logic and getting lost in the fog of intuition. Yet logical norms are blind to content and culture, ignoring evolved capacities and environmental structure. Often what looks like a reasoning error from a purely logical perspective turns out to be a highly intelligent social judgment in the real world. Good intuitions must go beyond the information given, and therefore, beyond logic.

A new scientific truth does not triumph by convincing its opponents and making them see the light, but rather because its opponents eventually die, and a new generation grows up that is familiar with it.

—Max Planck

Part 2

GUT FEELINGS IN ACTION

A good name is better than riches.

—Cervantes

7 ⋮ EVER HEARD OF . . . ?

The doorbell rang. The host rushed to the door to welcome the first guests arriving at the dinner party. He opened the door and turned to his wife. "May I introduce you to my new colleagues, Debbie and Robert." Then he turned to his guests. "And I would like to introduce you to my dear wife, umm, ah, umm . . ." Panic spread over his face until his wife helped him out. "Joanne," she said politely.

If a name sits too long on the tip of one's tongue, time can become painfully endless, particularly if the person in question is intimately related. But it could be worse. Not being able to *recall* a name occurs more often the older one gets, particularly if one has a Y chromosome. Yet if the husband no longer *recognized* his wife's name or her face, his gaffe would be in a different ballpark. He would be considered a clinical case and possibly end up in a neurological institution. Recognition memory is more reliable than recall memory both at the beginning and at the end of our lives and is also more fundamental; it is difficult, for instance, to recall individual information about a person whom one does not recognize.[1] Recognition memory is an evolved capacity that the recognition heuristic makes use of. We have already had a glimpse at recognition; let us take a closer look at it now.

No single rule of thumb can guide someone through an entire life. But the following story of a guy named Reese explores how mere recognition profoundly shapes our intuitions and emotions in everyday life.

Names, Names, Names

Reese was born in Spokane, a town in the state of Washington, where he spent his youth. He was recently nominated for a prize for outstanding emergency unit patient care and invited to fly to London. When he travels outside of the country, he envies people who can say that they are from New York, or from some other city that doesn't induce the question "Where?" He went to the Intercollegiate Center for Nursing Education in Spokane, near the Coeur d'Alene mineral fields. When he tells this, he is used to getting more blank stares and adds, "gold and silver mines," whereupon his listeners at least say, "Ah, yes." He rarely fails to add that Spokane was the place where, according to legend, Butch Cassidy and the Sundance Kid, the infamous bank and train robbers and foremost members of the Wild Bunch, died in obscurity. That usually gets him the response: "I've seen that movie. It's cool. But weren't they shot in Bolivia?" Whatever the truth is, at least these are names people recognize and they are linked to his hometown, which makes him feel an iota more important.

On the plane to London, Reese was seated next to a British woman in a Chanel suit who asked him what he did for a living. Reese explained that he works at the Westminster Clinic and hoped that she had heard of its reputation. In the course of their conversation he told her that he invested in stocks in order to pay for his children's college education, but that he would never buy any he hadn't heard of. He feels strongly about proper investing, because he wants his child to get a good education rather then ending up, as he did, at a no-name college. His three-year-old daughter already recognizes Mickey Mouse and Ronald McDonald. She loves to watch Disney movies and eat a Big Mac. Her face lights up when she hears of Madonna and Michael Jackson, even though she knows nothing about music. The little girl often begs him to buy toys advertised on TV and is afraid of people she has not seen before. The night before the trip, she got sick, and Reese drove her to a doctor they know instead of to a much closer clinic.

After he had arrived in London, he learned that the dinner invitation was black-tie. He did not own a tuxedo, nor did he recall a place to get one. When he looked up "London tailors" on the Internet, he recognized "Savile Row" and rushed there to buy a dinner jacket. At the dinner reception, he found himself standing in a large ballroom, uneasily scanning people dressed in black and white, looking in vain for a familiar face. He was relieved to see the woman from the plane. Reese was runner-up to the prize and had a good time in London until his last day, when one of his bags was grabbed by a thief. Asked for a description of the bag snatcher, he could not give reliable details, but later at the police station he recognized the man in one of the photos. Reese was glad to get back home to Spokane and be with his family. New faces and locations make him tense, but familiar ones give him a feeling of ease, even of intimacy.

RECOGNITION MEMORY

Recognition is the ability to tell the novel from the previously experienced, or the old from the new. Recognition and recall carve out our world into three states of memory. When visitors enter my office, they will be one of three kinds: those whose faces I do not recognize, those whom I recognize but don't recall anything else about (this is called *tartling* in Scotland), and those whom I both recognize and recall something about. Note that recognition memory is not perfect; I may have a wrong feeling of déjà vu, or not remember that I have already encountered a person. However, such errors need not be dysfunctional, because, as we will see shortly, forgetting can actually benefit the recognition heuristic.

The capacity for recognition is adapted to the structure of the environment. Herring gulls recognize their hatched chicks in order to rescue them from danger. Their nests are on the ground, making it easy for the chicks to stroll away and be killed by a neighbor. But they do not recognize their own eggs and are happy to sit on those of other gulls, or on a wooden dummy provided by an experimenter.[2] Outside of tricky experiments, they do not seem to need this capacity for recognizing eggs because their eggs don't roll far enough away to reach the next adult's nest. Lack of recognition can be exploited in the natural world as well. European cuckoos take advantage of other birds' inability to recognize their own eggs and offspring and lay their eggs in others' nests. The host birds seem to have wired into their brains the rule of thumb "feed any small bird sitting in your nest." In this particular bird environment, where nests are separated and chicks cannot move between nests, individual recognition is not necessary for taking care of the chicks.

In contrast, humans have an extraordinarily large capacity for recognizing faces, voices, and pictures. As we wander through a stream of sights, sounds, tastes, odors, and tactile impressions, some novel and some previously experienced, we have little trouble telling the two apart. In a remarkable experiment, participants were shown 10,000 pictures for five seconds each. Two days later, they correctly identified 8,300 of them.[3] No computer program to date can perform face recognition as well as a human child can. Why is this? As mentioned in chapter 4, humans are among the few species whose *unrelated* members exchange favors, such as trading goods, engaging in social contracts, or forming organizations. If we were not able to recognize faces, voices, or names, we would not be able to tell whom we'd encountered previously, and as a consequence, not recall who treated us fairly and who cheated. Hence, social contracts of reciprocity—"I share my food with you today, and you return the favor tomorrow"—could not be reinforced.

Recognition memory often remains when other types of memory, such as recall, become impaired. Elderly people suffering memory loss and patients suffering certain kinds of brain damage have problems saying what they know about an object or where they have encountered it. But they often know (or can act in a way that proves) that they have encountered the object before. Such was the case with R., a fifty-four-year-old policeman who developed such severe amnesia that he had difficulty identifying people he knew, even his wife and mother. One might be tempted to say he had lost his capacity for recognition. Yet in a test in which he was shown pairs of photographs consisting of one famous and one unfamous person, he could point to the famous persons as accurately as healthy people could.[4] It was his ability to recall anything about the persons he recognized that was impaired. Because recognition continues to operate even when everything else breaks down, I view it as a primordial psychological mechanism.

As the $1 million question in chapter 1 illustrates, the purpose of the recognition heuristic is not to recognize objects, but to make inferences about something else. Here we investigate it in more detail.

THE RECOGNITION HEURISTIC

The recognition heuristic is a simple tool from the adaptive toolbox that guides intuitive judgments, both inferences and personal choices. A judgment is called an *inference* when a single, clear-cut criterion exists, such as whether the Dow Jones will go up this week or whether or not a given player will win Wimbledon. Inferences can be right or wrong, and they can earn or lose a fortune. When no single, easily verifiable criterion exists, a judgment is called a *personal choice*—choosing a dress, a lifestyle, or a partner. Personal choices are more a matter of taste than of being objectively right and wrong, although the line between the two can get blurred.

Consider personal choices first. A business professor told me that he relies on brand-name recognition when purchasing a stereo system. He does not waste time consulting specialized magazines to learn about the avalanche of stereos featured on the market. Rather, he only considers brand names he has heard of, such as Sony. His rule of thumb is

> When you buy a stereo, choose a brand you recognize and the second-least expensive model.

Brand-name recognition narrows down choice and the principle of the second-lowest price is added to make the final decision. The rationale is that if one has heard of a company, it is likely because its products are good. The professor's justification for the additional step is that the quality of stereo technology has reached a level at which he is no longer able to hear the difference. His price principle arises from avoiding the cheapest and potentially least reliable model that companies manufacture for the low-price market. This rule saves time and likely protects him from being taken in.

Now consider inferences. The recognition heuristic can make

accurate inferences when there is a substantial correlation between recognition and what one wants to know. For simplicity, I assume that the correlation is positive. Here is the recognition heuristic for inferences about two alternatives:

> If you recognize one object but not the other, then infer that the recognized object has a higher value.

Whether the correlation is positive or negative can be learned from experience. Substantial correlations between name recognition and quality exist in competitive situations, such as the value of colleges, companies, or sports teams. When this correlation exists, ignorance is informative; the fact that you haven't heard of a college, firm, or team tells you something about it. An easy way to measure the degree to which your ignorance is informative is your *recognition validity*. Let's take the sixteen Gentlemen's tennis matches in the third round of Wimbledon 2003:

Andy Roddick—Tommy Robredo	Younes el Aynaoui—Andre Agassi
Roger Federer—Mardy Fish	Robin Soderling—Tim Henman
Paradorn Srichaphan—Rafael Nadal	Karol Kucera—David Nalbandian
Rainer Schuettler—Todd Martin	Radek Stepanek—Mark Philippoussis
Jonas Bjorkman—Justin Gimelstob	Sargis Sargsian—Juan Carlos Ferrero
Max Mirnyi—Ivo Karlovic	Jarkko Nieminen—Olivier Rochus
Feliciano Lopez—Flavio Saretta	Jiri Novak—Alexander Popp
Sjeng Schalken—Victor Hanescu	Wesley Moodie—Sebastien Grosjean

Neither an expert who recognizes all the names in a set nor someone totally ignorant of the players can use the recognition heuristic to infer the winner. Only if you are partially ignorant, that is, you have heard of some but not all of the players can this heuristic guide your intuitions. To determine your personal recognition validity, mark all the names of tennis players you have heard of. Now take the pairs in which you have heard of one player but not the other. Count all cases in which the player you recognized won the game (the winner is always listed first in the left-hand column, and second in the right-hand column). Divide this number by the number of all pairs in which you recognized one but not both, and you get your recognition validity for this set. For example, if you recognize only Roddick, Federer, Schuettler, Agassi, and Novak, the recognition heuristic predicts the outcomes right in 4 out of 5 times; that is, the validity is 80 percent. Relying on this rule of thumb, you would get only the Novak-Popp match wrong, although you know almost nothing. You can use your recognition most effectively if you have heard of half of the players, not more and not less. If your recognition validity is above 50 percent, there is wisdom in your ignorance; you do better than chance.

If you want to test it in other cases, here's the formula for all pairs in which one alternative is recognized and the other is not, in a class such as cities, companies, or sports teams:

> Recognition validity = number of correct inferences divided by number of correct plus incorrect inferences.

The recognition heuristic, just like every rule of thumb, does not always lead to the correct answer. As a consequence, the recognition validity is typically smaller than 100 percent.

Your recognition does not have the same validity for all problems, but depends on the class of objects (such as this year's Wimbledon Gentlemen's Singles contestants) and the kind of inference to be made (such as who will win). Consider potentially fatal diseases

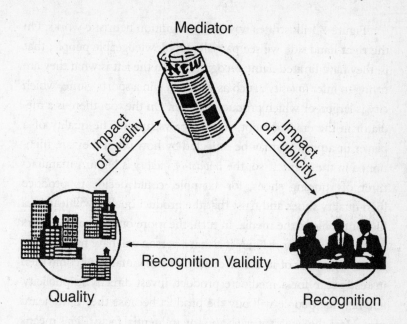

Figure 7-1: How the recognition heuristic works. *Impact of quality*: high-quality objects are mentioned more often in the media than low-quality objects. *Impact of publicity*: those that are mentioned more often are recognized more often. *Recognition validity*: those that are recognized more often are hence usually higher quality (based on Goldstein and Gigerenzer, 2002).

and infections. When people inferred which of two diseases, such as asthma or tularemia, is more frequent, a study reported recognition validities around 60 percent.[5] That is, the recognized disease was the more widespread one in 60 percent of the cases. This is better than chance, but not as good as for predicting the winners of Wimbledon tennis matches, which has been reported to be around 70 percent. When one wants to infer which of two foreign cities has a larger population, the validity is even higher, around 80 percent. In each case there exists a "mediator" that brings the names of diseases, players, or cities to the attention of the general public. These mechanisms include newspapers, radio, television, and word of mouth.

Figure 7-1 illustrates how the recognition heuristic works. On the right-hand side we see partially unknowledgeable people; that is, they have limited name recognition. On the left is what they are trying to infer (*quality*), such as who will win a sports game, which city is larger, or which product is better. On the top, there is a mediator in the environment, such as newspapers. The quality of a player or a product may be reflected by how often they are mentioned in the press. If so, the *impact of quality* is high. A manufacturer of running shoes, for example, could decide to produce high-quality shoes and trust that the product quality will lead to a higher profile in the media. In turn, the more often a name occurs in the news, the more likely it is that a person will have heard of the name, regardless of its actual quality. The manufacturer could then instead settle for a mediocre product, invest directly in publicity, and bet that people will buy the product because they have heard of it. Here the *impact of publicity* is an influential factor. This means shortcutting the triangle, as indeed many advertisers do. Measuring the impact of quality and publicity, we can predict in which situations relying on name recognition is informative or misleading.

So much for theory. But do people rely on the recognition heuristic in the real world? Let's begin with soccer matches.

ENGLISH FOOTBALL ASSOCIATION CUP

Founded in 1863, the Football Association (FA) is the ruling body for English football (soccer). It represents more than a million players belonging to tens of thousands of clubs and organizes national competitions. The FA Cup is the oldest soccer competition in the world and the major knockout tournament for English clubs. Teams are randomly paired, so that well-known clubs often compete against lesser-known clubs in lower divisions. Consider the following match from the third round of an FA Cup:

Manchester United plays *Shrewsbury Town*

Who will win? In a study, fifty-four British students and fifty Turkish students (living in Turkey) predicted the outcomes of this and thirty-one other FA Cup third-round matches.[6] British participants had plenty of knowledge about the previous records and current conditions of the two teams and could ponder the pros and cons before deducing who would win. The Turkish participants had very little knowledge about (or interest in) English soccer teams, and many protested their ignorance during the testing. Nevertheless, the Turkish forecasters were nearly as accurate as the English ones (63 percent versus 66 percent correct). The reason for this strong performance was that the Turkish laypeople intuitively followed the recognition heuristic consistently in 95 percent (627 out of 662) of all cases. Recall that an expert who has heard of all teams cannot use the recognition heuristic. A person who has never heard of Shrewsbury Town but only of Manchester United can guess the answer faster by relying on partial ignorance.

HOW COLLECTIVE WISDOM EMERGES FROM INDIVIDUAL IGNORANCE

Every year, millions of spectators watch the tennis matches at Wimbledon, one of the four annual grand slam tennis events, and the only one still played on natural grass. In the 2003 Gentlemen's Singles Championship, 128 players competed. We have already seen the names of the 32 players who made it into the third round. The players were ranked by the Association of Tennis Professionals and by the Wimbledon experts' seeding. For each of the 127 matches, one can predict that the player with the higher rank will win the game. In fact, the two ATP rankings predicted the winners correctly in 66 percent and 68 percent, respectively, of the matches. The experts did even slightly better. Their seeding predicted the outcome of 69 percent of the matches correctly.

How do ordinary people intuitively judge who wins? A study showed that in 90 percent of the cases where laypeople and amateur players had heard of one contestant but not of the other, they followed the recognition heuristic.[7] The amateurs had heard of the names of about half of the players, whereas the laypeople had only heard of fourteen, on average. All players were ranked according to the number of ordinary people who recognized their names, and the prediction was made that the one with higher name recognition would win. I refer to this ranking as *collective recognition*. Would you bet money on the combined ignorance of people who had not even heard of half of the Wimbledon contestants?

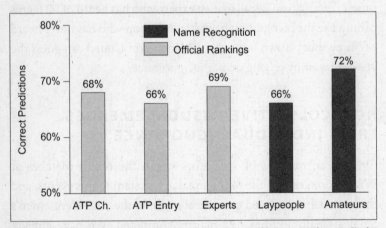

Figure 7-2: How to predict the outcomes of the 2003 Wimbledon's Gentlemen's Singles matches. Benchmarks are (1) the ATP Champions Race, the official worldwide ranking of tennis players for the calendar year, (2) the ATP Entry Ranking, the official ranking for the last fifty-two weeks, and (3) the seeding, which represents the expert ranking by Wimbledon officials. The outcomes were also predicted by the collective recognition of laypeople who had heard of only a few players and amateur players who recognized only half of them. Collective recognition based on partially ignorant people predicted the actual outcomes as well as or better than the three official benchmarks did (Serwe and Frings, 2006).

The collective recognition of the laypeople predicted the outcomes of 66 percent of the matches correctly, which was as good as the number predicted by the ATP Entry Ranking. That of the amateur group predicted the outcomes of 72 percent of the matches correctly, which was better than each of the three official rankings (Figure 7-2). A study of Wimbledon 2005 replicated this striking result. Both studies demonstrate that collective wisdom can emerge from individual ignorance and indicate that there is a beneficial degree of ignorance, where less knowledge is more. But they do not tell us when and why.

THE LESS-IS-MORE EFFECT

Let me begin with the odd story of how we discovered, or stumbled across, the less-is-more effect. We were testing a completely different theory for which we needed two sets of questions, one easy, the other difficult. For the easy set we chose one hundred questions—such as "Which city has more inhabitants, Munich or Dortmund?"—which were randomly drawn from information about the seventy-five largest German cities.[8] We asked students from the University of Salzburg, where I was teaching then, who knew lots about German cities. We thought a hundred similar questions from the seventy-five largest American cities would provide the difficult set, but when we saw the results, we could not believe our eyes (we had not yet read chapter 1 of this book). The students' answers were slightly more often correct for American cities, not German ones! I did not understand how people could answer questions equally well about something that they knew less about.

Salzburg has excellent restaurants. That night my research group had dinner at one of them to mourn the failed experiment. We tried in vain to make sense of this bewildering result. Finally, the insight dawned. If the students knew sufficiently little, that is, had not even heard of many of the American cities, they might intuitively rely on their ignorance as information. When it came to German cities, they were not able to do this. By exploiting the wisdom in missing knowledge, they scored equally well with the American cities. It has been said that researchers are like sleepwalkers whose creative intuition guides them to intellectual destinations they could never clearly see beforehand. Instead, I had been like a sleepwalker who had failed to understand the creative hunches of the intuitive mind. Discovery arose by serendipity: failing to do one thing and yet achieving another, more interesting one.

But how exactly does the less-is-more effect arise? Consider three American brothers who apply to a new school in Idaho. The principal tests each of them on their general knowledge, starting with geography. He names two European countries, Spain and Portugal, and asks which one has more inhabitants. The youngest brother goes first. He has not even heard of Europe, let alone the countries, and just guesses. The principal tests him on other pairs of countries, but the little brother performs at chance level and fails. Now it's the second brother's turn. Unlike his little brother, he occasionally watches TV news and has heard of half of the European countries. Even though he thinks he is guessing because he knows nothing specific about the countries he recognizes, he gets two-thirds of the questions correct and passes. Finally, the oldest brother is tested. He has heard of the names of all the countries, although he also knows nothing specific beyond their names. Surprisingly enough, he does worse than the second brother.

How can that happen? The youngest brother who has heard of none of the countries cannot use the recognition heuristic and performs at chance level (Figure 7-3). The oldest brother who has heard of all the countries cannot use it either and also performs at chance level (50 percent). Only the middle brother who has heard of some but not all of the countries can use the recognition heuris-

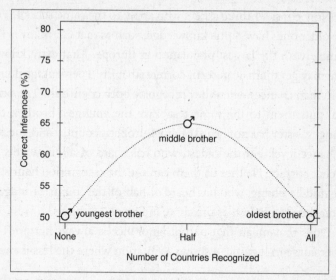

Figure 7-3: The less-is-more effect. Shown are three brothers who are asked which of two countries has more inhabitants. They know nothing about the countries, except that the middle brother has heard of the names of half of the countries, and the older has heard of all. The youngest brother and the oldest brother cannot rely on name recognition because they have heard of none and all, respectively, and perform at chance level. Only the middle brother can rely on name recognition, which improves his performance without knowing any facts.

tic; since he has heard of half of them, he can use the heuristic most often and does best. He scores 65 percent correct. Why? The recognition validity is 80 percent, a typical value. Half of the time the middle brother has to guess and half of the time he can use the heuristic. Guessing results in 25 percent correct (half of the half) and using the heuristic results in 40 percent correct (80 percent of the other half). Together, that makes 65 percent—much better than chance, even though he knows nothing about population sizes. The line in the figure that connects the three brothers shows how they would have performed at intermediate levels of name recognition. On the right side of this curve, a less-is-more effect is visible: the brother who recognizes all the countries performs less well.

Now consider three sisters who apply to the same school. The two older ones have some knowledge, such as that Germany is the country with the largest population in Europe. That extra knowledge may get them 60 percent correct (that is, 10 percentage points more than chance) when they recognize both countries.[9] The principal puts them to the same test. Like the youngest brother, the youngest sister has not heard of any European country and guesses at chance level, but the oldest, who has heard of all, now gets 60 percent correct. Neither of them can use the recognition heuristic. The middle sibling, who has heard of half of the countries, is again better than the oldest, as the curve in Figure 7-4 shows.

Does that mean that partial ignorance is always better? The upper curve in Figure 7-4 shows a situation where the less-is-more

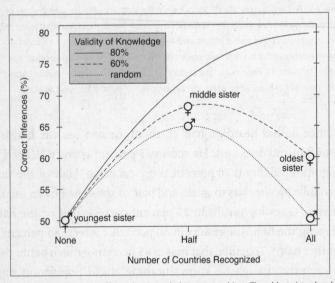

Figure 7-4: The less-is-more effect when people know something. The oldest sister has heard of all the countries and knows some facts and gets 60 percent of the questions right. The middle sister has not heard of half of the countries, and therefore can rely on name recognition and gets more answers right than her older sister. This less-is-more effect only disappears when the validity of knowledge is the same as that of recognition.

effect disappears. This is when the validity of one's knowledge matches or exceeds that of recognition. In the upper curve, both have a value of 80 percent. This means that someone knows enough to get 80 percent of the questions right when she recognizes both countries and 80 percent when she recognizes only one.[10] Here, less is no longer more.

Daniel Goldstein and I have shown that a less-is-more effect can emerge in different situations. First, it can occur between two groups of people, when a more knowledgeable group makes worse inferences than a less knowledgeable group—take the performance of the American and German students on the question of whether Detroit or Milwaukee is larger (chapter 1). Second, a less-is-more effect can occur between domains, that is, when the same group of people achieve higher accuracy in a domain in which they know little than in a domain in which they know a lot. For instance, when American students were tested on the largest American cities (such as New York versus Chicago) and on the largest German cities (such as Cologne versus Frankfurt), they scored a median 71 percent correct on their own cities but slightly higher on the less familiar German ones, with 73 percent correct.[11] This effect was obtained even though many Americans already knew the three largest U.S. cities in order and did not have to make any inferences. Third, a less-is-more effect can occur during knowledge acquisition, that is, when an individual's performance first increases but then decreases again. All these are expressions of the same general principle, which help to understand why tennis amateurs were able to make better predictions than the official rankings of the ATP and Wimbledon experts.

WHEN DOES FORGETTING HELP?

Common sense suggests that forgetting stands in the way of good judgment. Earlier in this book, however, we met the Russian mnemonist Shereshevsky, whose memory was so perfect that it was flooded with details, making it difficult for him to get the gist of a story. Psychologists have worked out in detail how forgetting can benefit the use of recognition.[12] Consider once again the oldest sister who recognizes all countries (Figure 7-4). If we could move her to the left on the curve toward the middle sister, she would do better. By forgetting some of the countries, she'd be able to use the recognition heuristic more often. Too much forgetting would be detrimental: if she moved too far to the left toward the youngest sister, she would again do worse.

In other words, if the oldest sister can no longer correctly remember every country she has heard of, this loss works to her advantage. This effect is only obtained if her memory errors are systematic, not random; that is, she tends to forget the smaller countries. Figure 7-4 shows that what constitutes a beneficial degree of forgetting also depends on the amount of one's knowledge: the more one knows, the less beneficial it is to forget. The oldest brother would gain more from forgetting than the oldest sister would. This would lead us to expect that people who know more about a subject forget less often. Did forgetting evolve to aid inferences by rules of thumb? We do not know. We are, however, beginning to understand that cognitive limits are not simply liabilities, but can enable good judgment.

WHEN IS IT SAFE TO FOLLOW THE MOST IGNORANT PERSON?

Let's look at the game show in chapter 1 again. This time the show master asks a group of three the $1 million question: "Which city has more inhabitants, Detroit or Milwaukee?" Again, none of them knows the answer for sure. If the members disagree on their best bet, one might expect that the majority determine the group decision. This is called the *majority rule*.[13] In an experiment, the following conflict arose. Two group members had heard of both cities and each concluded independently that Milwaukee was larger. But the third group member had not heard of Milwaukee, only of Detroit, and concluded that the latter was larger. What was their consensus? Given that two members had at least some knowledge about both cities, one might expect that the majority got their way. Surprisingly, in more than half of all cases (59 percent), the group voted for the most ignorant person's choice. This number rose to 76 percent when two members relied on mere recognition.[14]

It may seem odd that group members let their answers be dominated by the person who knew the least. But in fact one can prove that this is a successful intuition when the validity of recognition is larger than that of knowledge, which was the case for the participants in this experiment. Hence, the seemingly irrational decision to follow the most ignorant member increased the overall accuracy of the group. The study also showed a less-is-more effect in groups. When two groups had the same average recognition and knowledge validity, the group who recognized *fewer* cities typically had *more* correct answers. For instance, the members of one group recognized on average only 60 percent of the cities, and those in a second group 80 percent; but the first group got 83 out of 100 questions correct, whereas the second got only 75. Thus, group members seemed to intuitively trust the value of recognition, which can improve accuracy and lead to the counterintuitive less-is-more effect.

How conscious was the group decision to follow the wisdom in ignorance? The videotaped group discussions show that in a few cases, the least knowledgeable members articulated that a particular city must be smaller because they had not heard of it, and the others commented on that. Yet in most cases, the discussions lacked explicit verbalization or reasoning. What is noticeable, however, is that people who could rely on the recognition heuristic made snap decisions, which seemed to impress those who knew more and needed time to reflect.

SHOPPING FOR BRAND NAMES

If you read magazines or watch TV, you will have noticed that much advertisement is noninformative. The notorious Benetton campaign, for instance, only presented their brand name together with shocking images such as a corpse in a pool of blood or a dying AIDS patient. Why do firms invest in this type of advertisement? The answer is to increase brand-name recognition, important because of consumers' reliance on the recognition heuristic. Oliviero Toscani, the designer and man behind the Benetton campaign, pointed out that the ads had pushed Benetton beyond Chanel into the top-five best-known brand names across the world, and that Benetton's sales increased by a factor of ten.[15] If people did not rely on brand-name recognition in consumer choice, noninformative advertisement would be ineffective and so obsolete.

The effect of brand-name recognition extends to food as well. In an experiment, participants had a choice between three jars of peanut butter.[16] In a pretest, one brand had been rated as higher quality, and participants could identify the higher-quality product 59 percent of the time in a blind test (substantially higher than chance, which was 33 percent). With another group of participants, the scientists put labels on the jars. One was a well-known national brand that had been advertised heavily and which all participants recognized; the other two were brands they had never heard of before. Then the experimenters put the higher-quality peanut butter into one of the jars with the unfamiliar labels.

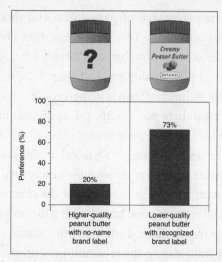

Figure 7-5: Brand names taste better. In a taste test, people were asked to compare high-quality peanut butter put in a no-name brand jar with low-quality peanut butter put in a jar from a nationally recognized brand (Hoyer and Brown, 1990).

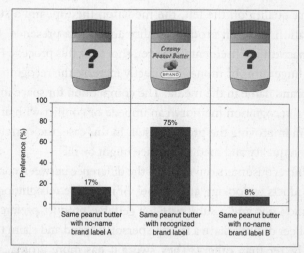

Figure 7-6: Brand-name recognition motivates consumer choice. In another taste test, the same peanut butter was in three different jars, one of which was a nationally recognized brand (Hoyer and Brown, 1990).

Would the same percentage of participants still choose the best-tasting peanut butter? No. This time 73 percent chose the low-quality product with the recognized brand label, and only 20 percent chose the high-quality product (Figure 7-5). Name recognition was more influential than taste perception. In a second tasting test, the researchers put exactly the same peanut butter into three jars, labeling two with unfamiliar labels and one with a brand-name label. The result was nearly identical. In this case, 75 percent of the participants chose the jar with the recognized brand, even though its content was the same as that of the other two jars (Figure 7-6). Nor did marking one brand with a higher price than the other two have much effect. Taste and price mattered little compared with the influence of the recognition heuristic.

Relying on brand-name recognition is reasonable when firms first increase product quality, and increased quality subsequently increases name recognition, by word of mouth or the media. This world is represented in Figure 7-1, with consumers on the right, product quality on the left, the media on the top, and a strong correlation between product quality and media presence. Non-informative advertisement, however, shortcuts this process. Firms spend huge sums of money to directly increase the recognition of their brand name in the media. The competition for space in consumers' recognition memory can impede or conflict with any interest in improving the product itself. In this case, the correlation between quality and media presence might be nil.

When consumers can only tell the difference between competing products by looking at the label, brand-name recognition and reputation become a substitute for genuine product preferences. Many beer drinkers have a favorite personal brand and claim that it tastes better than others. They swear it has more aroma, more body, is less bitter, and has just the right carbonation. These preferences are what some consumer theories take for granted, and that make more choice desirable as it gives every consumer a good chance of finding a match. Yet blind taste tests have repeatedly shown that consumers were unable to detect their own preferred brand. Some three hundred randomly selected American beer drinkers (who drank beer at least three times a week) were given five national and regional brands of beer.[17] The beer drinkers assigned "their" brands superior ratings over all competitors, as long as the label was on the bottle. When the label was taken off, and the test was "blind," none of the groups favoring a certain brand rated it as superior!

If consumers can only tell the difference between competing brands by name, then there is little economic justification for the idea that more choice is always better. Firms that spend their money on buying space in your recognition memory already know this. Similarly, politicians advertising their names and faces rather than their programs, and colleges, wannabe celebrities, and even small nations operate on the principle that if we do not recognize them, we will not favor them. Taken to the extreme, being recognized becomes the goal in itself.

DECIDING AGAINST NAME RECOGNITION

An effective use of the recognition heuristic depends on two processes, *recognition* and *evaluation*. The first asks, "Do I recognize this alternative?" and determines whether the heuristic can be applied. The second asks, "Should I rely on recognition?" and evaluates whether it should be applied to the situation at hand. For instance, most of us are hesitant to gather and eat unfamiliar mushrooms that we find while hiking in the woods. Yet when we find the same mushrooms on our plate in a good restaurant, we will probably not even think twice about gobbling them down. In the woods, we follow the recognition heuristic: if it's unknown, it might be poisonous. In the restaurant we do not follow it, because in this environment, the unknown is usually safe. This evaluation process is not always conscious. People intuitively "know" when a lack of recognition indicates a lack of safety.

The evaluation process is absent in automatic (reflex-like) rules of thumb. The recognition heuristic, by contrast, is flexible and can be consciously suppressed. How the evaluation process works is not yet known, but we have some clues. One aspect of this process seems to be whether reliable knowledge about what one wants to know can be retrieved. For instance, when members of Stanford University were asked whether Sausalito (a small town just north of the Golden Gate Bridge with only seventy-five hundred residents) or Heingjing (a made-up name that sounded like a Chinese city) had the larger population, most no longer relied on name recognition. They knew for sure that the town around the corner was small and guessed it must be Heingjing.[18] The source of recognition is another aspect that seems to be evaluated. In the same study, people were asked whether Chernobyl or Heingjing was larger, and only a few decided on Chernobyl, known mainly for its nuclear power plant accident, which has nothing to do with its size. Not relying on name recognition in such cases is an adaptive and smart response—except in this study, of course, where the experimenter deceived people by using a nonexistent city.

Is There a Neural Correlate of the Evaluation Process?

I have argued that the recognition heuristic is used in a flexible way. That means that the mind evaluates whether it should be used in a given situation. This kind of evaluation is what I refer to as the intelligence of the unconscious. If such an evaluation process exists, it should be separate from the processes of recognition. Therefore we should find distinct neural activities in the

Anterior frontomedian cortex

Figure 7-7: A neural correlate of the evaluation process. When people decide whether to rely on the recognition heuristic, a specific neural activity can be observed in the anterior fronto-median cortex (afMC).

brain when people decide whether to follow the heuristic. In a study using a brain imaging (functional magnetic resonance imaging, or fMRI) technique, we presented pairs of cities from Canada, England, France, Holland, Italy, Spain, and the United States to participants who were in a scanner.[19] One group's task was to infer which city in the pair had the larger population; the other group's task was simply to indicate which cities they had heard of. Note that the first task involves an evaluation process whereas the latter only involves recognition memory.

Did the brain scans indicate a neural correlate of this evaluation process? Such an activity needs to be very specific. That is, the particular brain activity should be observed when participants in the first group followed the recognition heuristic compared to when they did not, and should be absent when the participants in the second group simply indicated whether they recognized a city or not. Our study showed a specific activity in the anterior frontomedian cortex (Figure 7-7) of the first group but not the second. The location of the activity suggests that the process is not impulsive but a form of evaluation based on unconscious intelligence. There is still controversy about what this part of the brain actually does for us, though it has already been proposed that it serves evaluative functions, controls errors, and handles response conflicts. Its specific activity indicates that there is indeed a neural correlate to the evaluation process, that is, to the intelligence of the unconscious.

Where's My Steak?

No concert hall can afford a program that consistently ignores recognition. People like to hear the same music over and over again, and *La Traviata* fills the house that a nameless opera may leave vacant. This taste for familiarity can conflict with a taste for variety. In 2003, the Berlin Philharmonic toured the United States with Sir Simon Rattle as the conductor. At the time, the Berlin orchestra was arguably the best in the world. In New York, they performed Debussy's *La Mer*, which many did not know. In Rattle's words:

> During the New York performance of Debussy's *La Mer*, half of the audience sat there with folded arms and disapproving faces. They were there because they wanted a Porterhouse steak; instead we served them an unfamiliar meal. The other half of the audience perked its ears, but the first half remained skeptical to the end. "Where's my steak?" they asked themselves. It would be a mistake to give in and appease their culinary desires.[20]

Here, the audience mistrusted the musical taste of one of the best conductors in the world and went with the intuition "We don't know it, so we don't like it." Simon Rattle, who fills concert halls based on his name alone, can afford to ignore the audience's limited recognition of music, but less well known musicians and orchestras that strive for success rarely can.

There are more subtle ways to overcome novelty aversion. The eighteenth-century French economist and statesman Turgot was a reformer. As the story goes, he wanted to introduce potatoes to France, but peasants resisted the unknown food, until he came up with a trick. He ruled that only experimental state-run farms would be allowed to plant potatoes. Soon the peasants protested, clamoring for the privilege to grow them. Here, novelty aversion was trumped by a competitive social motive: "If another has it, I want it, too." Thus, the impulse to prefer what one recognizes can be overridden in various ways. A conductor like Rattle can ignore name recognition, although not consistently, and a politician like Turgot overcame it by putting the aversion to novelty in conflict with envy.

Nonetheless, much of the time the gut feeling to "go with what you know" is a helpful guide in life. An effective use depends on two processes, *recognition* and *evaluation*. The first decides whether the simple rule can be applied, the second whether it should. People tend to follow the recognition heuristic intuitively when it is valid, and collective wisdom based on individual ignorance can even outperform experts.

A man can be short and dumpy and getting bald
but if he has fire, women will like him.

—Mae West

8 : ONE GOOD REASON IS ENOUGH

Who would ever base an important decision on only one
reason? If anything unites the various tribes of rationality, it is the dictum that one should search for all relevant
information, weigh it, and add it up to reach a final judgment. Yet,
defying the official guidelines, people often base their intuitive
judgments on what I call *one-reason decision making*.[1] Keenly aware
of this tendency, so do many advertising campaigns. What did
McDonald's do when Burger King and Wendy's began to rival its
top-ranked name recognition? It launched a campaign that provided the *one* reason to choose McDonald's: "It's an easy way to
feel like a good parent." As an internal memo explained the underlying psychology, parents want their kids to love them, and
taking their children to McDonald's seems to accomplish this,
making them feel like good parents.[2] Wouldn't a few more reasons make a more convincing case? There is a proverb that a man
with too many good excuses shouldn't be trusted.

In this chapter, I will deal with intuitive judgments that are
based on recall memory. Recall goes beyond mere recognition; it
retrieves episodes, facts, or reasons from memory. I use the term
reasons to refer to cues or signals that help make decisions. Let's

first take a look at how evolution creates minds and social environments in which the use of *one good reason* spreads.

SEXUAL SELECTION

In most species of birds of paradise, the colorful males display, and the plain females choose. The males assemble in leks, that is, communal areas in which several of them stand in line or in groups and perform courtship displays, while the females stroll from candidate to candidate, scrutinizing them. How do the females decide on a mate? Most seem to rely on one reason only:

> Look over a sample of males, and go for the one with the longest tail.

Choosing a mate for only one reason may sound peculiar, but there are two theories to explain the practice.[3] The first is Darwin's theory of sexual selection elaborated by the statistician Sir Ronald A. Fisher. Females may have originally had a preference for a slightly longer tail because it allowed the male to fly and so get around better. If there is some genetic contribution to the natural variation in tail length, then a *runaway process* toward longer tails can be set in motion. Every female that deviated from the rule and chose a small-tailed male would be penalized, because if she did not produce long-tailed sons, her sons would have a smaller chance of being regarded as attractive and being able to reproduce. Tails accordingly grew longer over generations and eventually became widely accepted as attractive by females. Thus, the process of sexual selection can produce one-reason decision making in the minds of animals and an environment of long tails, bright colors, and other extravagant secondary sexual characteristics.

Darwin considered two mechanisms underlying dazzling male

features. The first was male-male competition, which had led to developments such as deer antlers and antelope horns. But battles between males could not account for the train of the peacock, so Darwin proposed a second mechanism, the power of female choice. He believed that females have a sense of beauty and are excited by the extravagant ornaments displayed by males. Darwin's theory of sexual selection was almost completely ignored for nearly a hundred years.[4] His male contemporaries could not be convinced by the idea that birds or deer would have a sense of beauty, and even less, that female taste could influence the evolution of male bodily features. Men close to Darwin, such as Thomas Henry Huxley, known as Darwin's bulldog, tried to persuade him to abandon the theory of sexual selection.[5] Today, this theory is a vigorously pursued branch of biology, and one might speculate whether its acceptance has been facilitated by the rising public role of women in Western societies. Yet we are only beginning to understand the connection between sexual selection and one-reason decision making. Just as the theory of sexual selection was rejected for a long time in biology, so the theory that one good reason is a viable strategy is still controversial in decision theory. But there is hope, as science itself is evolving.

Handicaps

The second theory that accounts for the tail of the bird of paradise and similar features is Amotz Zahavi's *handicap principle*. Whereas in the runaway theory of sexual selection the male may or may not be of good quality (and once the tail size escalates, the female is no longer choosing on grounds of quality), according to the handicap principle, he actually *is* of good quality. Birds of paradise's tails, just like those of peacocks, evolved precisely because they are handicaps. A male bird shows off his tail because it advertises that he can survive in spite of it. In this theory, the one good reason—the big handicap—is truly a good one. In the runaway selection theory, the one good reason is a deceptive one, an initially good reason that grew out of control. Despite their differences in interpretation, both theories explain how one-reason decision making can spread.

The handicap principle was also rejected outright by the scientific community. Not until 1990 did this verdict change.[6] Around that time, experiments with peacocks (who also assemble in leks, where they display their ornamented trains to visiting females) indicated that peahens also base their choice on one reason. The only factor correlated with mating success was the number of eyespots a male had on his feathers. But this correlation could be due to other factors; strong males might have more eyespots. If males were left the same in every other way but had fewer eyespots, would the females no longer choose them? In an ingenious experiment, British researchers cut away 20 of around 150 eyespots from half of the males under study, and handled the other half in the same stressful way without actually removing the spots. They reported a sharp decline in the mating success of those with a reduced number of eyespots compared to that of the previous season and to those who still had their full number of spots. Moreover, the researchers did not observe a single peahen that mated with the first male that courted her; rather, peahens sampled on average three males before they chose. In almost all cases, the peahen chose the male with the highest number of eyespots in her sample.[7] It is likely that the peahen's one good reason is genetically coded.

Thus, both sexual selection and the handicap principle can produce one-reason decisions in both minds and their environments. The simultaneous evolution of genes (coding decision rules) and environment is called coevolution. We might find the minimalism of the bird of paradise's mate choice amusing. However, it seems to have worked for millennia, and in fact, seems to exist in humans. The one good reason is often social, such as when a woman desires a man and primarily falls in love with him because he is desired by other women. This one reason virtually guarantees that a woman's peer group will accept and admire the choice she has made.

Irresistible Cues

The eyespots on a peacock's tail are a powerful cue for the peahen. In general, environments are populated by more or less irresistible cues that control animal, including human, behavior. As already mentioned, some cuckoos leave their eggs in the nests of other birds, which hatch and feed the cuckoo chicks. In one species, a single irresistible cue, a patch on the cuckoo chick's wing that flutters and simulates the gaping mouths of many hungry chicks, fools the foster parents into feeding them. The behavior of the host birds is sometimes cited as evidence that they cannot discriminate between the cuckoo chicks and their own because of cognitive limitations. But a little girl cuddling baby dolls has no problem in distinguishing between them and human babies; the dolls' cuteness simply elicits her maternal instincts. Similarly, a man who squanders his time looking at porn magazines is captivated by the photos of nude women though he knows they're not the real thing.

Irresistible cues can be the product of cultural transmission as well as of evolution. Voting is a case in point. The scheme of political Left-Right is a simple cultural cue that provides many of us with an emotional guide for what is right and wrong in politics. It is so emotionally overwhelming that it can also structure what is politically acceptable in our everyday lives. People who think of themselves as politically left-wing may not want to be friends or even talk with someone who is politically right-wing. Similarly, for some conservatives, a socialist or communist is almost an alien life-form. Let us have a closer look at this overpowering cue that shapes our identity.

THE ONE-DIMENSIONAL VOTER

With the collapse of the Soviet Union, democracy has become universally praised as the preeminent form of government in Europe and North America. Its institutions guarantee us advantages that our grandmothers and grandfathers were willing to risk their lives for: freedom of speech, freedom of the press, equality of citizens, and constitutional assurances of due process, among others. Yet there is a paradox. Philip Converse's seminal study, *The Nature of Belief Systems in Mass Publics*, revealed that American citizens as a rule are badly informed about political choices, have not thought through the issues, and can be easily blown from one side of an issue to the other.[8] It's not that people know nothing; it's just that they know nothing about politics. The most widely known fact about George H. W. Bush in the presidential elections of 1992 was that he hated broccoli. And almost all Americans knew that the Bushes' dog was named Millie, while only 15 percent knew that both Bush and Clinton favored the death penalty.[9] Converse was not the first to notice this shocking degree of ignorance. The existence of chronic, often opinionated *know-nothings* was noted in Europe as well. Karl Marx spoke of the *lumpenproletariat*, people he felt were easily propagandized, manipulated, and mobilized to act against the interests of the working classes. Marx was explicit about what he thought of the makeup of this bottom stratum of society:

> Vagabonds, discharged soldiers, discharged jailbirds, escaped galley slaves, swindlers, mountebanks, lazzaroni, pickpockets, tricksters, gamblers, procurers, brothel keepers, porters, literati, organ grinders, rag pickers, knife grinders, tinkers, beggars—in short, the whole indefinite, disintegrated mass, thrown hither and thither.[10]

More than a century later, in the 1978 gubernatorial race in Georgia, the candidate Nick Belluso broadcast a television ad. The candidate's consultants seemed to think the American public's opinion was just as moldable as that of Marx's lumpenproletariat. Here is the commercial:

Candidate: This is Nick Belluso. In the next ten seconds you will be hit with a tremendously hypnotic force. You may wish to turn away. Without further ado let me introduce to you the hypnogenecist of mass hypnosis, the Reverend James G. Masters. Take us away, James.

Hypnotist (in strange garb, surrounded by mist): Do not be afraid. I am placing the name of Nick Belluso in your subconscious mind. You will remember this. You will vote on Election Day. You will vote Nick Belluso for governor. You will remember this. You will vote on Election Day. You will vote Nick Belluso for governor.[11]

Perhaps because most TV stations refused to run the ad (some fearing the effects of hypnosis on the viewer), the ad was a flop; Belluso lost the vote and went on to run unsuccessfully for a number of positions, including president in 1980. Many political ads are equally edifying, if less amusing, when it comes to the candidate. Few ads in modern democracies provide information about the issues; most rely on increasing name recognition by means of repetition, on creating negative emotions toward the opponent, or simply on one-liners, laughter, and the politics of entertainment. How can citizens have an opinion about parties if they know so little about them? In deference to Herbert Simon, this mystery is known as *Simon's puzzle*.[12] It is the paradox of mass politics.

Left-Right Redux

In 1980, something unique took place in the history of German democracy. A new party, the Greens, competed in the federal election, challenging the established system. This event marked the beginning of a swift career from a citizens' initiative against nuclear power to a partnership in the federal government by the end of the twentieth century. A successful new party is a rare affair and poses a new twist to Simon's puzzle: how can citizens have an opinion about a new party when they barely know anything about the old ones?

Let's first look at the time before the Greens emerged. Back then, six parties occupied the German political landscape. Scores of issues divided them: religious versus secular orientation, economic policies, social welfare orientation, family and immigrant politics, and moral issues, including abortion. Citizens knew about most of these issues, but their preferences showed no such complexity. Rather, most voters' preferences for the six parties were based on only one reason: where the parties fall in the continuum of Right and Left. Voters perceived the parties like six pearls on a string (Figure 8-1). This string of pearls has served as a model for political life in France, Italy, the United Kingdom, and the United States as well. In the United States, where the political Left in the European sense is almost nonexistent, it has also been called *liberal versus conservative*. Voters agree where the parties are located in the one-dimensional landscape, but they disagree on which ones they like and hate.

Helga Q. Public's "ideal point" on the string is close to her favorite party, say, the Liberals. Can one predict how she ranks the other parties? Yes, and quite simply. Helga "picks up" the string at her ideal point, which leaves both ends hanging down in parallel strings.[13] She does not have to know anything about the other parties

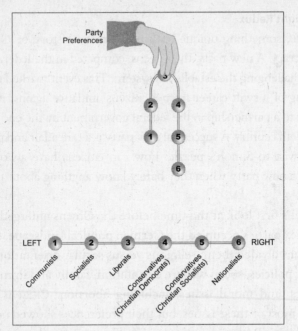

Figure 8-1: The string heuristic. Voters tend to reduce the complexity of the political landscape to one dimension: Left-Right. Parties are mentally arranged like pearls on a string. By picking up the string at one's ideal point, a voter can "read off" his or her preferences for the other parties. Shown are six political parties in Germany (Socialists=Social Democratic Party; Liberals=Free Democratic Party; Nationalists=National Democratic Party).

beyond Left-Right, but she can nevertheless "read off" her preferences for a new (or old) party by relying on what I call the *string heuristic*:

> The closer a party is to your ideal point on the continuum of Left-Right, the higher your preference.

The string heuristic determines the party preferences within each of the two "ends" in Figure 8-1. For instance, Helga Q. prefers the Socialists to the Communists, the Christian Democrats to the

Christian Socialists, and both to the Nationalists. If she preferred the Communists to the Socialists, or the Nationalists to the Christian Socialists, each of these preferences would contradict the hypothesis that she uses the string heuristic. How many voters actually rely on it? For the classical six-party system, 92 percent of all voters I studied followed this simple rule of thumb.[14] This is how consistent preferences are formed, even when the voter is fairly uninformed.

How did voters react to the new party? The platform of the Greens did not easily fit into the old Left-Right scheme. The Greens brought together issues such as the protection of the forest and the closing down of nuclear power plants, which united both conservative forest workers and left-wing intellectuals who feared worldwide radioactive pollution following a nuclear meltdown. Would the party be assimilated into the old Left-Right scheme, or would the scheme be extended to new issues, such as environmentalism? To answer this question, I studied a group of 150 university student voters. Among them, 37 percent voted for the Greens. Although voters located the Greens all over the Left-Right continuum, each individual voter's preferences were consistent and stable over time. Their party preferences continued to follow the string heuristic, only now the string had one more pearl. But what happened to the ecological dimension? From all we know, ecological orientation has little to do with the Left or Right, as it has been a key issue for neither. In fact, voters' perception of how ecologically oriented a party was could be derived from its orientation toward the Left or Right. When one picked up the string at the point where a voter had located the Greens, the two ends revealed this voter's rankings of the parties' ecological orientation. Yet none of the voters appeared to be conscious of its mechanisms.

Einstein is reported to have said, "Politics is more difficult than physics."[15] That may be true, but to the degree we understand the cognitive processes involved, Simon's puzzle can be resolved, piece by piece. The string heuristic explains how politically uninformed people get a feeling for where parties stand on issues and allows these voters to form consistent opinions. In a two-party system, such as in the United States, this mechanism is even simpler. When is the string heuristic reasonable? It seems to work in systems where political institutions present themselves along the Left-Right divide, arranging and polarizing their issues accordingly. An issue that is initially only loosely associated with a party, such as a stance in favor of abortion or against the death penalty, will become more and more attached to it as political opponents take the opposite stance. When this happens, voters can indeed read off from Left to Right where parties stand on the issues, even when the stance is often little more than historical accident. Consistent with this hypothesis, political campaigns and media reports use the Left-Right vocabulary, and political scientists construct their research instruments accordingly.[16] Wherever the string heuristic and party politics coevolve in this way, the heuristic becomes helpful. The one-dimensional voter is able to "know" where parties stand without actually knowing.

SEQUENTIAL DECISIONS

Both the peacock's tail and the political Left-Right scheme are irresistible cues. Yet one single cue is not sufficient for all situations. There is another class of intuitive judgments in which one or more cues are looked up in memory, although again only one of them determines the final decision. A process in which one first considers one cue, and, if it does not allow a decision, considers another, and so forth, is called sequential decision making. Put yourself in the following situation.

Parents' Nightmare

Around midnight, your child is short of breath, coughing and wheezing. You desperately look for an available doctor. In your phone book there are two entries for local after-hours primary health care. One is a general practitioner who would come to your home within twenty minutes. You know him from your health center, and he never listens to what you have to say. The other entry is an emergency center sixty minutes away, run by general practitioners. You don't know these doctors, but you've heard that they listen to parents. Whom do you call? And why?

In the United Kingdom, parents of children under thirteen were asked this and similar questions.[17] The pairs of consultations described were variations of four reasons that earlier studies had identified to be of primary concern for British parents: where their child was seen, by whom, the time taken between call and treatment (waiting time), and whether the doctor listened to them. Many seemed to carefully weigh each of the four reasons and combine them into their decision. However, almost every second parent had one dominant reason that made their decision. For the largest group, over a thousand parents, it was whether or not the doctor listened to what they were saying— even if it meant waiting forty minutes longer. These parents were more likely to be female, well educated, and have more children. For about 350 parents, the dominant reason was waiting time. Fifty parents needed to see a physician they knew, whether or not the doctor listened and no matter how long the waiting time would be. The fourth reason—whether the child was seen at home or in an emergency center—by contrast, was not a dominant reason for any of the parents.

How can we understand these parents' intuitions? Assume that the order of importance is from top to bottom: the doctor's tendency to listen, waiting time, familiarity, and location. Now consider the choice between primary health care A and B, available in one night:

	A or B?	
Does the doctor listen?	Yes	No

STOP & PICK A

Since the first reason already allows for a decision, the search for further information is stopped, all other potential reasons are ignored, and the parents go for alternative A. On another night, the choice may be more complicated:

	C or D?	
Does the doctor listen?	Yes	Yes
How long to wait?	20 min	20 min
Do you know the doctor?	No	Yes

STOP & PICK D

In the second situation, neither the first nor the second reason allows for a decision. The third one does, and the parents therefore go for alternative D, using the Take the Best heuristic, which we encountered in the context of predicting dropout rates in high schools. It consists of three building blocks:

Search rule: Look up reasons in the order of importance.
Stopping rule: Stop search as soon as the alternatives for one reason differ.
Decision rule: Choose the alternative that this reason suggests.

This process is also described as lexicographic—when one looks up words in a lexicon, one has to look up the first letter first, then the second, and so on. Dozens of experimental studies have shown that people's judgments tend to follow Take the Best, and identified conditions where this likely happens.[18] Intuitions that rely on Take the Best may need to search through several reasons, but finally rely on just one to make the decision.

To now, we have seen how parents made this important decision, but we do not know how good their preferences would be. Many authorities on rational decision making would be appalled to hear how these parents dealt with such a life-and-death issue, given their low esteem of lexicographic rules such as Take the Best:

> We examine an approach that we believe is more widely adopted in practice than it deserves to be: lexicographical ordering. However, it is simple and it can easily be administered. Our objection is that it is naively simple. . . . Again, we feel that such an ordering procedure, if carefully scrutinized, will rarely pass a test of "reasonableness."[19]

This message is from two eminent figures in rational decision making, who seemed so sure about their verdict that they did not bother to conduct the test themselves. In order to test how "reasonable" sequential decision making is, we need to look at a situation where a clear-cut outcome exists. What would be better than sports?

Take the Best

Over one thousand matches were played in the National Basketball Association's (NBA's) 1996–97 season. Undergraduate students from New York University were asked to predict which team would win in a random sample of all games from that season. They were given only two clues: the number of games won in the season (the base rate), and the score at halftime of the game. In order to prevent other information from influencing the predictions, the names of the teams were not given. In more than 80 percent of the cases, the intuitive judgments were found to be consistent with Take the Best. Here is the theory of how it worked. The first clue was the number of games won. If both teams differed by more than fifteen games, then search was stopped and the person guessed that the team with the higher amount would win that game. The following NBA match is an illustration.

	A or B?	
Games won	60	39

STOP & PICK A

Since the first clue allowed for a decision, the information about the halftime result was ignored, and the prediction was that team A would win. If the difference in games won was smaller than fifteen, the second clue, the score at halftime, was considered:

	C or D?	
Games won	60	50
Halftime score	36	40

STOP & PICK D

Since team D was ahead at halftime (here the difference in scores didn't matter), the prediction was that it would win.

But how accurate are intuitions based on only one reason? As mentioned above, according to traditional theories of rationality, these intuitions are doomed to failure. One should never ignore reasons, but combine number of games won and halftime score. In this view, hunches that rely on Take the Best commit one of two "sins." If, as in the first example, an intuitive judgment relies solely on the base rate information (the number of games won) and ignores the halftime score, it commits what has been called *conservatism*. Conservatism means that only the old information is taken into consideration and the new information, the halftime result, is ignored. If, as in the second example, intuition is instead based on the halftime score only, this "sin" has been labeled the *base-rate fallacy*.[20]

These alleged sins are presented in virtually every psychology textbook as verdicts against intuition: people may use simple rules as a shortcut, but it's naive to do so. Yet, as I discussed earlier, Take the Best can predict school dropout rates faster and more accurately than a complex version of Franklin's rule does. The NBA study provided another test of Take the Best, this time against Bayes's rule, the Goliath of rational strategies.[21] Bayes's rule does not waste information. It always uses base rates and halftime score, as well as the actual difference in scores at halftime, whereas Take the Best either ignores halftime or simply considers who is ahead in a match. The question is, if people followed the rational Bayes's rule, how much more accurate would they be in predicting the outcomes of all 1,187 NBA games than if their intuitions followed the simple Take the Best heuristic?

The test showed by way of computer simulation that if people used Bayes's rule, they would be able to predict 78 percent of the winners correctly. Take the Best, in spite of committing alleged sins against rationality, predicted exactly the same percentage of winners, yet more quickly and with less information and computation.

Something must be wrong, it seems, with this result. Yet it held in soccer as well. A student of mine replicated the result with soccer matches in the German major league, the Bundesliga, for the seasons 1998–2000. Take the Best used the same order of reasons. For both seasons, it predicted the outcomes of more than four hundred games as well as or better than Bayes's rule did.[22] The advantage of the simple rule was most pronounced when the base rates (the games won) were two years old rather than from the previous season, that is, when the problem was most difficult. In each case, Take the Best embodied the intuition that if one team had been considerably more successful than its opponent in a previous season, it would likely win again; otherwise, the team that led at halftime would win. The complex calculations could not beat this intuition.

When Is One Good Reason Better Than Many?

The idea that intuitions based on Take the Best could be as accurate as complex decision making is hard to swallow. When I presented the first results to international groups of experts, I asked them to estimate how closely Take the Best would match the accuracy of a sophisticated modern version of Franklin's rule (multiple regression). *Not a single one* expected that the simple rule could be as accurate, let alone that it could be better, and most estimated that Take the Best would fail by 5 to 10 percentage points or more. To their surprise, across twenty studies, using Take the Best produced more accurate decisions. Since then, we have shown the power of one good reason in a broad range of real-world situations.[23]

These results are important in demonstrating that intuitions based on one good reason not only are efficient but can also be highly accurate. In fact, even if a mind were able to compute the most sophisticated artificial intelligence strategies available today, it would not always do better. The lesson is to trust your intuition when thinking about things that are difficult to predict and when there is little information.

Recall that in an uncertain world, a complex strategy can fail exactly because it explains too much in hindsight. Only part of the information is valuable for the future. A simple rule that focuses only on the best reason and ignores the rest has a good chance of hitting at the most useful information.

Imagine a plot with 365 points representing the daily temperature in New York for one year. The values are low in January, rise through spring and summer, and then decrease again. The pattern is quite jagged. If you are mathematically sophisticated, you can find a complex curve that fits the points almost perfectly. But this curve is not likely to predict next year's temperatures as well. After the fact, a good fit is, by itself, worth little. By trying to find the perfect fit, one fits irrelevant effects that do not generalize to the future. A simpler curve will give better predictions for next year's temperature, even though it does not match the existing data as well. Figure 5-2 illustrates exactly this principle. The more complex strategy is better than the simple one in hindsight, but not in predictions. In general:

> *Intuitions based on only one good reason tend to be accurate when one has to predict the future (or some unknown present state of affairs), when the future is difficult to foresee, and when one has only limited information. They are also more efficient in using time and information. Complex analysis, by contrast, pays when one has to explain the past, when the future is highly predictable, or when there are large amounts of information.*[24]

DESIGNING OUR WORLD

Evolution appears to have designed various animal minds to rely on sequential cue assessment.[25] For instance, female sage grouse first assess males in a lek on the basis of their songs and visit only those who pass this first test for a closer inspection. Such a sequential process of mate choice seems to be widespread and has also been observed in food choice and navigation. Honeybees trained in experiments to identify model flowers rely on a set order of cues, with odor on the top. They decide on the basis of color only if the odor of two flowers matches and on the basis of shape only if odor and color match. Yet the cue that is consulted first is not always the most valid one; in some cases the order is determined by how far a sense can reach out into the world. In an environment in which trees and bushes block vision, acoustic cues tend to be available before visual and other cues. For instance, a deer stag estimates the strength of a rival first by the deepness of its roar, and only later by vision. If these two reasons were not enough to make it run off scared, it will experience the most authoritative signals of its opponent's strength when they begin fighting.

Sequential decisions based on one good reason, however, are not animals' only adaptive strategies. Adding and averaging of two or more cues seem to occur with individual differences between young and old, experienced and inexperienced. For instance, older female garter snakes appear to demand males that are good on two cues, whereas either cue alone satisfies younger snakes.

As we have seen before, intuition, like evolution, takes advantage of one good reason. We humans can also consciously take advantage of it in designing our world. Sequential decisions can make the environment safer, more transparent, and less confusing.

Contest Rules

Playing in the World Cup tournament is the ultimate dream of every national soccer team. In the first round, groups of four teams compete. The two best teams in each group move on to the next round. But how to determine who is the "best"? The FIFA (International Federation of Football Associations) considers six aspects of performance to be relevant.

1. Total points in all matches (three for a win, one for a tie)
2. Points in the direct competition matches
3. Difference in number of goals in the direct competitions
4. Number of goals in the direct competitions
5. Difference in number of goals in all matches
6. Number of goals in all matches

Let's look first at the ideal of making trade-offs, that is, weighing and adding. A committee of international experts could come up with a weighing scheme. For instance, one could give the first aspect a weight factor of 6, the second a weight of 5, and so on. That, one might argue, would be fairer and allow a more comprehensive evaluation of performance than would simply judging teams on one of these aspects. Designing such a scheme, however, clearly invites endless discussion; the American Bowl Championship Series formula that ranks college football teams by complex weighing and adding sparked a wave of complaints. An equally worrisome problem is that a weighing scheme is not intuitively transparent. Coaches, players, reporters, and fans alike would be busier calculating the final score results than enjoying the game.

The alternative is to dispense with weighing and adding, and to introduce Take the Best. That is what the FIFA uses. The aspects of performance are ordered as above, and if two teams differ on the first, the decision is made. Only when the teams are tied is the second aspect looked up, and so on. Everything else is ignored. Sequential decisions based on one good reason can be easily made and embody transparent justice.

Safety Design

Trade-offs may spoil the fun when it comes to ball games, but in other contexts would be downright dangerous. When determining which car has the right of way on a crossing, there are several potentially relevant factors.

1. The hand signals of the police officer regulating the traffic
2. The color of the traffic light
3. The traffic sign
4. Where the other car comes from (right or left)
5. Whether the other car is bigger
6. Whether the driver of the other car is older and deserves respect

Imagine a world in which traffic laws give each of these factors due consideration, because it is considered fair to make trade-offs between public signs and courtesy. Yet weighing and counting pros and cons would be unsafe, since drivers do not have the time and might make computational errors. Two cars may be close in size, and the age of the other driver may be difficult to assess. A safer design, and the one used in all the countries I know of, is sequential one-reason decision making. If there is a police officer who regulates the traffic, all other signals in the above list must be ignored by a driver approaching the crossing. If there is no police officer, only the traffic light counts. If there is no traffic light, then the traffic signs are decisive. One could of course imagine an alternative system in which all that counts is whether the other car is bigger, but this is fortunately not yet law and only practiced in isolated cases.

Traffic laws that make trade-offs would have a different structure. For instance, the police officer's order to stop could be overruled if both a green traffic light and the traffic sign indicated otherwise. Or a green traffic light would be overruled if the traffic sign was a yield sign and the other car was bigger. Continually making trade-offs would turn our everyday world into a risky place by slowing down decisions that need to be made fast.

Number Design

You enter a party and see a bunch of people. How many are there? An adult with no special training has a direct perception of up to only four people. That is, one immediately knows how many others are in the room, if they do not number more than four. Beyond that, humans have to count. This psychological capacity of four has become a building block of various cultural systems. Romans, for instance, gave ordinary names to the first four of their sons, but the fifth and every subsequent one was named by counting, that is, by a numeral: Quintus, Sixtus, Septimus, and so on. Similarly, in the original Roman calendar, the first four months had names, Martius, Aprilis, Maius, Junius; the fifth and all others were referred to by their order number: Quintilis, Sextilis, September, October, November, December.[26]

Various present-day cultures in Oceania, Asia, and Africa only have the words *one*, *two*, and *many*. But that does not mean that they cannot do arithmetic. People have designed various systems to count. Some cultures use wooden sticks to keep a tally; others tick things off with the parts of their bodies, matching numbers to a sequence of fingers, toes, elbows, knees, eyes, nose, and so on. Tally notches and marks have been found on animal bones and cave walls, probably twenty to thirty thousand years old. The tally system is the source of the Roman numbering system, where I is one, II is two, III is three, V is short for five, X for ten, C for hundred, D for five hundred, and M for thousand. Like the ancient Greek and Egyptian systems, the Roman numerals made calculation a pain. These cultures were locked up for centuries with tallying systems that were incoherent and unusable for most purposes except writing a number down.

The breakthrough came from Indian civilization, which gave us our modern "Arabic" system. Its genius lies in its introduction of a lexicographic system, which is inherent in the sequential rules discussed in this chapter. Take a quick look at the following two numbers, represented in the Roman system. Which one is larger?

MCMXI
MDCCCLXXX

Now look at these two numbers represented in the Arabic system.

1911
1880

One can see immediately that the first number is larger when represented in the Arabic, but not in the Roman system. Roman numerals represent magnitude neither by the length of the numeral nor by its order. In terms of length, MDCCCLXXX should be larger, but is not. In terms of order, from left to right, after the M (representing a thousand) there is a C (representing a hundred) in MCMXI, whereas there is a D (representing five hundred) in MDCCCLXXX. Nevertheless, the first number is larger. But the Arabic system is based strictly on order. If two numbers have the same length as in the example, one only needs to search from left to right for the first digit that is different. One can then stop searching and conclude that the number with the higher digit is the larger number. All other digits can be ignored. Putting order into our representations of the world can generate insight in our minds and simplify our lives.

It is ironic, moreover, that the best lessons in "fast and frugal rules of thumb" may well come from understanding the cognitive processes of those master clinicians who consistently make superb decisions without obvious recourse to the canon of evidence-based medicine.

—C. D. Naylor[1]

9 : LESS IS MORE IN HEALTH CARE

A glass of red wine at dinner prevents heart attacks; butter kills you; all treatments and tests are desirable, as long as you can afford them—most of us have strong intuitions about what is good and bad in health care. Although we act on these beliefs, they are typically based on rumor, hearsay, or trust. Few make a serious effort to find out what medical research knows, although many consult consumer reports when buying a refrigerator or computer. How do economists make health care decisions? We asked 133 male economists at the 2006 meeting of the American Economic Association whether they take PSA (prostate-specific antigen) tests for prostate cancer screening, and why. Among those over fifty, the majority participated in screening, but very few had read any medical literature on the topic and two-thirds said that they did not weigh the pros and cons of screening.[2] Most just did whatever their doctor told them to do. Like John Q. Public, they relied on the gut feeling:

> If you see a white coat, trust it.

Trust in authority, rumor, and hearsay were efficient guides in human history before the advent of books and medical research. Learning by firsthand experience was potentially deadly; finding out by oneself which plants were poisonous was a bad strategy. Is blind trust in the health expert still sufficient today, or do patients need to research more carefully? The answer depends not only on the expertise of your doctor but on the legal and financial system in which your health care system operates.

CAN DOCTORS TRUST PATIENTS?

Daniel Merenstein, a family physician, is not sure he will ever be the doctor he wants to be. As a third-year resident, he saw a highly educated fifty-three-year-old man for physical examination.[3] They discussed the importance of diet, exercise, wearing seat belts, and the risks and benefits of screening for prostate cancer. While proper diet, exercise, and seat belts have proved beneficial to health, there is no proof that men who participate in screening with PSA tests live longer than those who don't—contrary to what some physicians and patients believe. But there is proof that those who test positive may be harmed through treatments for slow-growing cancers that, even if untreated, would not cause problems in a man's lifetime. After radical prostatectomy about three out of ten men can become incontinent and six out of ten become impotent.[4] This is why nearly all national guidelines recommend that physicians discuss the pros and cons of PSA tests with the patient, and why the U.S. Preventive Services Task Force concludes that the evidence is insufficient to recommend for or against routine PSA screening.[5] Merenstein spent much time keeping up-to-date with current medical studies so he could practice what is known as evidence-based medicine. After learning about the pros and cons, the patient declined the PSA test. Merenstein never saw the man again, and after he graduated the patient went to another office. His new doctor ordered PSA testing without discussing with him its risks and benefits.

The patient was unlucky. He was subsequently diagnosed with a horrible, incurable form of prostate cancer. Although there is no evidence that early detection of this cancer could have saved or prolonged the man's life, Dr. Merenstein and his residency were put to trial in 2003. Merenstein assumed that he'd be accused of failing to discuss prostate cancer screening with the patient. Yet the plaintiff's attorney claimed that the PSA test was the standard of care in the Commonwealth of Virginia and that Merenstein should have ordered the test, not discussed it. Four Virginia physicians testified that they simply do the test without informing their patients. The defense brought in national experts who testified that the benefits of PSA screening are unproved and questionable, whereas severe harms are documented, and emphasized the national guidelines of shared decision making.

In his closing arguments, the plaintiff's lawyer contemptuously referred to "evidence-based medicine" as merely a cost-saving method, naming the residency and Merenstein as its disciples and the experts as its founders. He called upon the jury to return a verdict that would teach residencies not to send more doctors out on the streets believing in evidence-based medicine. The jury was convinced. Merenstein was exonerated, but his residency was found liable for $1 million. Before the trial, Merenstein believed in the value of keeping up with the current medical literature and bringing it to the patient. He now looks at the patient as a potential plaintiff. Being burned once, he feels he has no choice but to overtreat patients, even at the risk of causing unnecessary harm, in order to protect himself from them. "I order more tests now, am more nervous around patients; I am not the doctor I should be."[6]

CAN PATIENTS TRUST DOCTORS?

The story of young Kevin in the second chapter makes us wonder about the damage caused by overdiagnosis in health care. Merenstein and his residency have learned the hard way that they are supposed to perform tests on their patients in order to protect themselves, even if a test's potential harms are proved and its potential benefits are not. Clearly something is going wrong with health care. The good old-fashioned gut feeling "If you see a white coat, trust it" has done much good. But it cannot work as well when physicians fear lawsuits, overmedication and overdiagnosis have become a lucrative business, and aggressive direct-to-consumer advertising has become legal. All lead instead to a decrease in the quality and an increase in the costs of health care. Let me define two consequences:[7]

> *Overdiagnosis* is the detection of a medical condition through testing that otherwise would not have been noticed within the patient's lifetime.
> *Overtreatment* is the treatment of a medical condition that otherwise would not have been noticed within the patient's lifetime.

Would you rather receive a thousand dollars in cash or a free total-body computed tomography (CT) scan? In a telephone survey of a random sample of five hundred Americans, 73 percent said they would prefer the CT.[8] Do these optimists know what they are getting? Obviously not. There is no evidence to support the benefit or even safety of a total-body CT screening; it is not endorsed by any professional medical organization, and even discouraged by several.[9] Nonetheless, CT scans and other high-technology screening tests are successfully marketed by an increasing number of independent entrepreneurs, including physicians. Professional TV actors dressed up as doctors spread slogans like "Take the test, not the chance."

Physicians who sell CT screening might respond that people have the right to make use of it without waiting years before its effectiveness or harms have been proved—after all, a normal result can give consumers "peace of mind." Sounds comforting, but is it true that one has peace of mind if the CT results are normal? Absolutely not; it's more an illusion of certainty. Consider electron-beam CT, which is performed to identify persons with an increased risk of coronary artery disease. The chance that it correctly identifies persons with increased risk is only 80 percent; that is, 20 percent of those who are at risk are sent home with a false peace of mind. Its false-alarm rate is even worse. Among people who are *not* at risk, 60 percent are nevertheless told that their results are suspicious.[10] That is, many of those who have no reason to worry may spend the rest of their lives frightened about a nonexistent medical condition. I have rarely heard of such a poor high-tech test, worse than other noninvasive and less expensive testing methods. I myself would rather pay a thousand dollars to avoid the test—and save my peace of mind.

Do doctors take the tests they recommend to patients? I once gave a lecture to a group of sixty physicians, including representatives of physicians' organizations and health insurance companies.

The atmosphere was casual, and the organizer's warm personality helped to develop a sense of common agenda. Our discussion turned to breast cancer screening, in which some 75 percent percent of American women over fifty participate. A gynecologist remarked that after a mammogram, it is she, the physician, who is reassured: "I fear not recommending a mammogram to a woman who may later come back with breast cancer and ask me 'Why didn't you do a mammogram?' So I recommend that each of my patients be screened. Yet I believe that mammography screening should not be recommended. But I have no choice. I think this medical system is perfidious, and it makes me nervous."[11] Another doctor asked her whether she herself participates in mammography screening? "No," she said, "I don't." The organizer then asked all sixty physicians the same question (for men: "If you were a woman, would you participate?"). The result was an eye-opener: not a single female doctor in this group participated in screening, and no male physician said he would do so if he were a woman.

If a woman is a lawyer, or the wife of a lawyer, does she get better treatment? Lawyers seem to be regarded by doctors as especially litigious patients who should be treated with caution when it comes to risky procedures such as surgery. The rate of hysterectomy in the general population in Switzerland was 16 percent, whereas among lawyers' wives it was only 8 percent—among female doctors it was 10 percent.[12] In general, the less well educated a woman is and the better private insurance she has, the more likely it is that she'll get a hysterectomy. Similarly, children in the general population had significantly more tonsillectomies than the children of physicians and lawyers. Lawyers and their children apparently get better treatment, but here, better means less.

So what do you do if your mother is sick and you want to know what your doctor really thinks? Here is a helpful rule:

> Don't ask your doctors what they recommend. Ask them what they would do if it were their mother.

My experience has been that doctors change their advice when I ask about their mother or other relatives. The question shifts their point of view; a mother would not sue. Yet not every patient is ready to accept that doctors are under external pressure, and that patients must therefore take on some responsibility for their treatment. The doctor-patient relationship is deeply emotional, as the case of a friend and novelist illustrates.

"We can't meet tomorrow morning, I've got to go to my doctor," he told me.
"I hope it's nothing serious?"
"Only a colonoscopy," my friend reassured me.
"Only? Do you have pain?"
"No," he replied, "my doctor said I need to have one, I'm forty-five. Don't worry, in my family, nobody ever had colon cancer."
"It can hurt. Did your doctor tell you what the possible benefits of a colonoscopy are?"
"No," my friend said, "he just said it's a routine test, recommended by medical organizations."
"Why don't we find out on the Internet?"

We first looked up the report of the U.S. Preventive Services Task Force. It said that there is insufficient evidence for or against routine screening with colonoscopy. My friend is Canadian and responded that he does not bank on everything American. So we looked up the Canadian Task Force report, and it had the same result. Just to be sure, we checked Bandolier at Oxford University in the United Kingdom, and once again we found the same result. No serious health association we looked up reported that people should have a routine colonoscopy—after all, a colonoscopy can be extremely unpleasant—but many recommended the simpler, cheaper, and noninvasive fecal occult blood test. What did my friend do? If you think that he canceled his doctor's appointment the next day, you are as wrong as I was. Unable to bear the evidence, he got up and left, refusing to discuss the issue any further. He wanted to trust his doctor.

DOCTORS' DILEMMA

Patients tend to trust their doctors, but they do not always consider the situation in which the doctors find themselves. Most physicians try to do their best in a world in which time and knowledge are severely limited. In the United States, the average time patients have to describe their complaints before they are interrupted by their physicians is twenty-two seconds. The total time the physician spends with a patient is five minutes—"how are you" and other formal niceties included. That is markedly different in countries such as Switzerland and Belgium that have an "open market" in which patients have access to more than one general practitioner or specialist. In this competitive situation the doctor invests time in his patients to encourage them to return. Here the average duration of a visit is fifteen minutes.[13]

Continuing education is indispensable in the rapidly changing world of medicine. Yet most physicians have neither the time to read even a few of the thousands of articles published every month in medical journals nor the methodological skills to evaluate the claims in these articles. Rather, continuing education mostly happens in seminars sponsored by the pharmaceutical industry, usually at a nice vacation spot, with spouses' and other expenses included. Pharmaceutical firms conveniently provide summaries of scientific studies of their featured products, which their representatives distribute in the form of advertisements and leaflets to physicians. As a recent investigation revealed, these are not neutral summaries. The assertions in 175 different leaflets distributed to German physicians could be verified in only 8 percent of the cases.[14] In the remaining 92 percent of cases, statements in the original study were falsely reported, severe side effects of medication were not revealed, the period during which medication could safely be taken was exaggerated, or—should doctors have wanted to check the original studies—the cited source was not provided or was impossible to find. As a consequence, many physicians have only a tenuous connection with the latest medical research.

For patients and doctors alike, geography is destiny. The surgeons in one medical referral region in Vermont removed the tonsils of 8 percent of the children living there, while those in another community removed the tonsils of 70 percent of the children. In one region in Iowa, 15 percent of all men had undergone prostate surgery by age eighty-five; in another region, it was 60 percent. Women are subject to this same geographical power over their bodies. In one region in Maine, 20 percent of the women had a hysterectomy by the age of seventy; in another region, over 70 percent underwent this operation.[15] There is little reason to believe that these striking regional differences have much to do with patients' conditions. Whether or not people undergo a treatment depends on local custom, while the kind of treatment depends on the attending physician. For localized prostate cancer, for instance, most urologists recommend radical surgery, whereas most radiation oncologists recommend radiation treatment. The authors of the Dartmouth Atlas of Health Care conclude that "the 'system' of care in the United States is not a system at all, but a largely unplanned and irrational sprawl of resources, undisciplined by the laws of supply and demand."[16]

At a time when everyone is worried about exploding health care costs, we spend billions of dollars every year on care that provides little or no benefits to people, and sometimes even causes them harm. Can we counteract these problems and instill in our health care system a good dose of rationality? The system in fact needs a three-pronged cure: it must develop efficient and transparent policies in place of physicians' defensive practices and local custom; it needs to find common ground between medical experts about good treatment; and finally it needs reform in the practice of litigation that allows physicians to do what is best for the patient rather than follow self-protecting procedures. In the next section, I illustrate how one can achieve the first goal.

HOW TO IMPROVE PHYSICIANS' JUDGMENTS

There are two classical proposals, both of which follow the spirit of Franklin's rule. According to clinical-decision theory, patients and doctors should choose between alternative treatments by surveying all possible consequences, and then estimating the numerical probability and utility of each consequence. One then multiplies these, adds them up, and chooses the treatment with the highest expected utility. The beauty of this approach is that it embodies shared decision making: the physician provides the alternatives, consequences, and the probabilities, and the patient is responsible for attaching numbers to the potential benefits and harms. Yet decision theorists have convinced few doctors to engage in this calculation because it is time-consuming, and most patients resist attaching numerical values to the potential harms of a tumor versus those of a heart attack. Proponents of clinical-decision analysis will respond that their intuitions need to be changed, yet proof that expected-utility calculations are the best form of clinical decision does not exist, and there are even reports that they do not always lead to better decisions. Last but not least, when intuition clashes with their deliberate reasoning, people tend to be less satisfied with the choice they make.[17]

The second proposal is to introduce complex statistical aids for physicians making treatment decisions, which might lead to better results than do their intuitions.[18] We'll see this type of method in the next section. Although these decision aids are more widely adopted than expected-utility calculations, they are still rare in clinical practice, and again at odds with medical intuition. A majority of physicians don't understand complex decision aids and end up abandoning them. As a result, physicians are left with their own clinical intuition biased by self-protective treatment, specialty, and geography.

Is there a way to respect the nature of intuitions *and* improve treatment decisions? I believe that the science of intuition provides such an alternative. To that end, I was glad to read in the renowned medical journal *Lancet* that our research on rules of thumb is starting to have an impact on medicine. As the epigraph to this chapter reveals, rules of thumb are seen as an explication of clinical masterminds' intuitions. Yet in the same issue of the *Lancet*, another article provided a different interpretation of our work: "The next frontier will involve fast and frugal heuristics; rules for patients and clinicians alike."[19] Here, rules of thumb are seen as an alternative to complex decision analysis. My own conviction is that physicians already use simple rules of thumb but for fear of lawsuits do not always admit it. Instead they tend to use these rules either unknowingly or covertly, leaving them little possibility for systematic learning. The ensuing problems for health care are obvious. My alternative is to develop intuitive decisions into a science, discuss them openly, connect them with the available evidence, and then train medical students to use them in a disciplined and informed way.

The following story illustrates that program. It looks at three ways to make treatment allocations: by clinical intuition, by a complex statistical system, and by a fast and frugal rule of thumb. The story begins several years ago, when I gave a talk to the Society for Medical Decision Making in beautiful Tempe, Arizona. I explained in what situations simple rules can be faster, less costly, and more accurate than complex strategies. When I stepped down from the podium, Lee Green, a medical researcher from the University of Michigan, approached me and said, "Now I think I understand my puzzle." Here is his story.

TO THE INTENSIVE CARE UNIT?

A man is rushed to a hospital with severe chest pains. The emergency physicians suspect the possibility of a heart attack (acute ischemic heart disease). They need to act, and quickly. Should the man be assigned to the coronary care unit or to a regular nursing bed with electrocardiographic telemetry? This is a routine situation. Every year, between one and two million patients are admitted to coronary care units in the United States.[20] How do doctors make this decision?

In a Michigan hospital, doctors relied on the long-term-risk factors of coronary artery disease, including family history, being male, advanced age, smoking, diabetes mellitus, increased serum cholesterol, and hypertension. These physicians sent about 90 percent of the patients with severe chest pain into the coronary care unit. This is a sign of defensive decision making; doctors fear being sued if patients assigned to a regular bed die of a heart attack. As a consequence, the care unit became overcrowded, the quality of care decreased, and costs went up. You might think that even if a patient doesn't have a heart attack, it was better to be safe than sorry. But being in the ICU carries its own risks. Some twenty thousand Americans die every year from a hospital-transmitted infection, and many more contract one. Such infections are particularly prevalent in the intensive care unit, making it one of the most dangerous places in the hospital—a dear friend of mine died in the ICU from a disease he'd picked up there. Yet when putting patients into this extremely dangerous situation, doctors protect themselves from being sued.

A team of medical researchers from the University of Michigan was called in to improve conditions. When they checked the quality of physicians' decisions—and quality control is not yet always the rule in hospitals—they found a disturbing result. Not only did doctors send most patients into the unit; they sent those who should have been there (who had a heart attack) as often as those who should not have been there (who did not have a heart attack). Doctors' decisions were no better than chance, but nobody seemed to notice. As a second study revealed, the long-term-risk factors doctors were looking for were not the most relevant ones for discriminating between patients with and without acute ischemic heart disease. Specifically, the physicians looked for a history of hypertension and diabetes, "pseudo-diagnostic" cues, instead of the nature and location of patients' symptoms and certain clues in the electrocardiogram, all of which are more powerful predictors of a heart attack.[21]

What to do? The team first tried to solve the complex problem with a complex strategy. They introduced the *heart disease predictive instrument*.[22] It consists of a chart with some fifty probabilities and a long formula that enable the physician, with the help of a pocket

		Chest Pain = Chief Complaint EKG (ST, T wave Δ's)				
History	ST&T Ø	ST⇔	T⇑⇓	ST⇔	ST⇔&T⇑⇓	ST⇑⇓&T⇑⇓
No MI & No NTG	19%	35%	42%	54%	62%	78%
MI or NTG	27%	46%	53%	64%	73%	85%
MI and NTG	37%	58%	65%	75%	80%	90%

		Chest Pain NOT Chief Complaint EKG (ST, T wave Δ's)				
History	ST&T Ø	ST⇔	T⇑⇓	ST⇔	ST⇔&T⇑⇓	ST⇑⇓&T⇑⇓
No MI & No NTG	10%	21%	26%	36%	45%	64%
MI or NTG	16%	29%	36%	48%	56%	74%
MI and NTG	22%	40%	47%	59%	67%	82%

		No Chest Pain EKG (ST, T wave Δ's)				
History	ST&T Ø	ST⇔	T⇑⇓	ST⇔	ST⇔&T⇑⇓	ST⇑⇓&T⇑⇓
No MI & No NTG	4%	9%	12%	17%	23%	39%
MI or NTG	6%	14%	17%	25%	32%	51%
MI and NTG	10%	20%	25%	35%	43%	62%

Figure 9-1: The heart disease predictive instrument chart. It comes with a pocket calculator. If you don't understand it, then you know why most physicians don't like it.

calculator, to compute the probability that a patient should be admitted to the coronary care unit. The physicians were taught to find the right probabilities for each patient, type these into the calculator, press ENTER, and read off the resulting number. If it was higher than a given threshold, the patient was sent to the care unit. A quick glance at the chart makes it clear why the physicians were not happy using this and similar systems (Figure 9-1). They don't understand them.

Nevertheless, after the physicians were first exposed to the system, their decisions improved markedly and overcrowding eased in the coronary care unit. So the team surmised that calculation, rather than intuition, worked in their cases. But they were well-trained researchers and tested their conclusion by taking the chart and the pocket computer away from the physicians. If calculation were the key, the quality of their decisions should fall back to the initial chance level. Yet the physicians' performance did not drop. The researchers were surprised. Had the physicians memorized the probabilities on the chart? A test showed that they hadn't, nor had they understood the formula in the pocket calculator. The researchers then returned the calculator and the chart to the physicians, withdrew them again, and so on. It made no difference. After the physicians' first exposure to the chart, their intuitions improved permanently, even without further access to the calculating tools. Here is the puzzle: how could the physicians make the right calculations when they no longer had the key tools?

It was at this point that I met Green, the principal investigator, and it was during my talk that he found the answer: the physicians did not need the chart and the calculator because they did not calculate. But what then improved their intuitions? All that seemed to matter were the right cues, which the physicians had memorized. They still worked with their intuitions, but now they knew what to look for, whereas earlier they had looked in the wrong places. This insight opened up a third alternative, beyond mere intuition and complex calculation, a rule of thumb for coronary care allocations, designed by Green together with David Mehr. It corresponded to the natural thinking of physicians but was empirically informed. Let me explain the logic of constructing such a rule.

Transparent Diagnostic Rules

The heart disease predictive instrument was proved effective on some twenty-eight hundred patients in six New England hospitals. Why not use it in another hospital, such as in Michigan? As I mentioned before, it lacks transparency. When systems with heavy calculation and scores of probabilities conflict with their intuitions, physicians tend to avoid the more complicated method.[23] Yet there is another drawback to complexity that we saw in the last chapter. When there is high uncertainty, simple diagnostic methods tend to be more accurate. Predicting heart attacks is extremely difficult, and no even remotely perfect method exists.

Let us take for granted that the predictive instrument is excellent for the New England patients, but it does not necessarily follow that it will perform equally well in Michigan. The patients in the Michigan hospitals differ from those in New England, but we do not know how and to what extent. One way to find out would be to start a new study with several thousand patients in the Michigan hospitals. That option is not available, however, and even if it were, such a study would take years. In the absence of data, we can use the simplifying principles introduced in the previous chapters.

But how? One way is to reduce the number of factors in the complex diagnostic instrument, and use one-reason decision making. That would lead to a *fast and frugal tree* (see below). It is like Take the Best but can solve a different class of problems: classifying one object (or person) into two or more categories.

Fast and Frugal Tree

A fast and frugal tree asks only a few yes-or-no questions and allows for a decision after each one.[24] In the tree developed by Green and Mehr (Figure 9-2), if there is a certain anomaly in the electrocardiogram (the so-called ST segment), the patient is immediately admitted to the coronary care unit. No other information is required. If that is not the case, a second cue is considered: whether the patient's chief complaint was chest pain. If not, the patient is assigned to a regular nursing bed. All other information is ignored. If the answer is yes, then a final question is asked. This third question is a composite one: whether any of the other five factors is present. If so, the patient is sent to the coronary care unit. This decision tree is fast and frugal in several respects. It ignores all fifty probabilities and all but one or a few diagnostic questions.

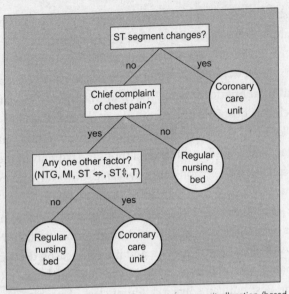

Figure 9-2: A fast and frugal decision tree for coronary care unit allocation (based on Green and Mehr, 1997).

This fast and frugal tree puts the most important factor on the top. Changes in the ST segment send the endangered patients quickly into the care unit. The second factor, chest pain, sends patients who shouldn't be in the care unit to a regular nursing bed in order to reduce dangerous overcrowding. If neither of these factors is decisive, the third one comes into play. Physicians prefer this fast and frugal tree to a complex system, because it is transparent and can be easily taught.

But how accurate is such a simple rule? If you were rushed to the hospital with severe chest pains, would you prefer to be diagnosed by a few yes-or-no questions or by the chart with probabilities and the pocket calculator? Or would you simply trust a physician's intuitions? Figure 9-3 shows the diagnostic accuracy of each of these three methods in the Michigan hospital. Recall that

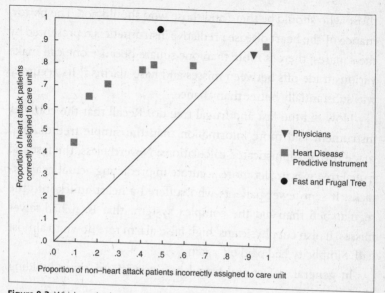

Figure 9-3: Which method can best predict heart attacks? The three methods shown are physicians' intuitive judgments, the complex heart disease predictive instrument, and the fast and frugal tree.

there are two aspects to accuracy. On the vertical axis is the proportion of patients correctly assigned to the coronary care unit (i.e., who actually had a heart attack), which should ideally be high; on the horizontal axis is the proportion of patients incorrectly assigned, which should be low. The diagonal line represents performance at the level of chance. Points above the diagonal show performance better than chance, and those below reflect performance worse than chance. A perfect strategy would be in the upper-left-hand corner, but nothing like this exists in the uncertain world of heart disease. Physicians' accuracy before the intervention of the Michigan researchers was at chance level—even slightly below. As mentioned before, they sent about 90 percent of the patients into the care unit but could not discriminate between those who should be there and those who should not. The performance of the heart disease predictive instrument is represented by the squares; there is more than one square because one can make various trade-offs between misses and false alarms.[25] Its accuracy was substantially better than chance.

How did the fast and frugal tree do? Recall that the complex instrument had more information than the simple tree did and made use of sophisticated calculations. Nevertheless, the fast and frugal tree was in fact more accurate in predicting actual heart attacks. It sent fewer patients who suffered a heart attack into the regular bed than did the complex system; that is, it had fewer misses. It also cut physicians' high false-alarm rate down to almost half. Simplicity had paid off again.[26]

In general, a fast and frugal tree consists of three building blocks:

Search rule: Look up factors in order of importance.
Stopping rule: Stop the search if a factor allows it.
Decision rule: Classify the object according to this factor.

A fast and frugal tree is different from a full decision tree. Full trees are not rules of thumb; they are information-greedy and complex rather than simple and transparent. Figure 9-4 shows both kinds of trees. A full tree has 2^n exits or leaves, whereas a fast and frugal tree has only $n + 1$ (where n is the number of factors). When looking at four factors, this makes 16 versus 5 leaves (see Figure 9-4). With 20 factors, this makes 1,000,000 versus 21 leaves. Constructing full trees runs into other problems as well. Not only do they quickly become computationally intractable, but as a tree grows in size, there are less and less data available to provide reliable estimates for what to do at each stage. For example, if you start with ten thousand patients and try to divide them

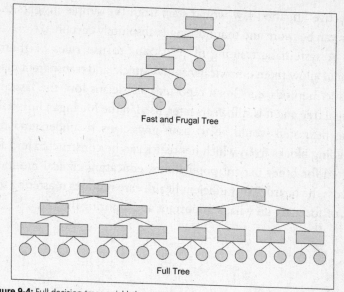

Fast and Frugal Tree

Full Tree

Figure 9-4: Full decision trees quickly become computationally intractable when the number of cues increases, whereas fast and frugal trees do not.

up among the million leaves, you will end up with unreliable information. Unlike the full tree, the fast and frugal tree introduces order—which of the factors are the most important ones?—to make itself efficient.

Medical Intuition Can Be Trained

The moral of the overcrowding story is this: physicians' intuitions can be improved not only by complex procedures that are in danger of being misunderstood and avoided, but by simple and empirically informed rules. The latter can reduce overcrowding, increase the quality of care, and decrease the wide variability in physicians' treatment choices. Geography no longer need be destiny, and physicians no longer need to make unreliable decisions. Yet this change in methodology must be supported by legal reform that frees physicians from a fear of doing the best for their patients. An effective litigation law would start from the simple insights that less can be more and that nothing is absolutely certain.

A systematic training of physicians to use rules of thumb would allow them empirically sound, quick, and transparent diagnostic methods. As Green reported, physicians love the fast and frugal tree and it is still, years later, used in the Michigan hospitals. The next step would be to train physicians to understand the building blocks from which heuristics can be constructed and adjusted for other patient populations, educating clinical intuition across the board. Truly efficient health care requires mastering the art of focusing on what's important and ignoring the rest.

> There is nothing divine about morality; it is a purely human affair.

> —Albert Einstein

10 : **MORAL BEHAVIOR**

ORDINARY MEN

On July 13, 1942, the men of the German Reserve Police Battalion 101, stationed in Poland, were awakened at the crack of dawn and driven to the outskirts of a small village. Armed with additional ammunition, but with no idea what to expect, the five hundred men gathered around their well-liked commander, the fifty-three-year-old major Wilhelm Trapp. Nervously, Trapp explained that he and his men had been assigned a frightfully unpleasant task and that the orders came from the highest authorities. There were some eighteen hundred Jews in the village who were said to be involved with the partisans. The order was to take the male Jews of working age to a work camp. The women, children, and elderly were to be shot on the spot. As he spoke, Trapp had tears in his eyes and visibly fought to control himself. He and his men had never before been confronted with such an order. Concluding his speech, Trapp made an extraordinary offer: if any of the older men did not feel up to the task that lay before them, *they could step out.*

Trapp paused for a moment. The men had a few seconds to

decide. A dozen men stepped forward. The others went on to participate in the massacre. Many of them, after they had done their duty once, vomited or had other visceral reactions that made it impossible to continue killing and were then assigned to other tasks. Almost every man was horrified and disgusted by what he was doing. Yet why did only a mere dozen men out of five hundred declare themselves unwilling to participate in the mass murder?

In his seminal book *Ordinary Men*, historian Christopher Browning describes his search for an answer, based on the documents from the legal prosecution of the Reserve Police Battalion 101 after the war. There were detailed testimonies of some 125 men, many of which "had a 'feel' of candor and frankness conspicuously absent from the exculpatory, alibi-laden, and mendacious testimony so often encountered in such court records."[1] An obvious explanation would be anti-Semitism. Yet Browning concludes that this is unlikely. Most of the battalion members were middle-aged family men, considered too old to be drafted into the German army and conscripted instead into the police battalion. Their formative years had taken place in the pre-Nazi era, and they knew different political standards and moral norms. They came from the city of Hamburg, by reputation one of the least nazified cities in Germany, and from a social class that had been anti-Nazi in its political culture. These men did not seem to be a potential group of mass murderers.

Browning examines a second explanation: conformity with authority. But the extensive court interviews indicate that this was not the primary reason either. Unlike in the Milgram experiment, where an authoritative researcher told participants to apply electric shocks to other people, Major Trapp explicitly allowed for "disobedience." His extraordinary intervention relieved the individual policemen from direct pressure to obey the order from the highest authorities. The men who stepped out experienced no sanctions from him, although Trapp did have to restrain a captain who was furious that the first man to refuse duty was from his company. If it was neither anti-Semitism nor fear of authority, what had turned ordinary men into mass killers? Browning points to several possible causes, including the lack of forewarning and time to think, concern about career advancement, and fear of retribution from other officers. Yet he concludes that there is a different explanation, based on how men in uniforms identify with their comrades. Many policemen seemed to follow a social rule of thumb:

Don't break ranks.

In Browning's words, the men felt "the strong urge not to separate themselves from the group by stepping out"[2] even if conforming meant violating the moral imperative "don't kill innocent people." Stepping out meant losing face by admitting weakness and leaving one's comrades to do more than their share of the ugly task. For most, it was easier to shoot than to break ranks. Browning ends his book with a disturbing question: "Within virtually every social collective, the peer group exerts tremendous pressures on behavior and sets moral norms. If the men of Reserve Police Battalion 101 could become killers under such circumstances, what group of men cannot?" From a moral point of view, nothing can justify this behavior. Social rules, however, can help us understand why certain situations promote or inhibit morally significant actions.

ORGAN DONORS

Since 1995, some fifty thousand U.S. citizens have died waiting in vain for a suitable organ donor. As a consequence, a black market in kidneys and other organs has emerged as an illegal alternative.

Although most Americans say they approve of organ donation and in most states it is possible to register online, relatively few have actually signed a donor card. Why are only 28 percent of Americans but a striking 99.9 percent of French citizens potential donors?[3] What keeps Americans from signing and saving lives?

If moral behavior were the result of deliberate reasoning, then the problem might be that Americans are not aware of the need for organs. That would call for an information campaign to raise public awareness. Dozens of such campaigns have already been launched in the United States and in other countries yet have failed to change the consent rate. But the French apparently don't need to enlighten their citizens. One might speculate about national characters. Have the French reached a higher stage of moral development, or are they less anxious than the Americans about having their bodies opened postmortem? Perhaps Americans fear that, as several popular novels and films have suggested, emergency room doctors won't work as hard to save patients who have agreed to donate their organs. But why are only 12 percent of Germans potential donors, compared to 99.9 percent of Austrians? After all, Germans and Austrians share language and culture and are close neighbors. A glance at the striking differences in Figure 10-1 shows that something very powerful must be at work, something that is stronger than deliberate reasoning, national stereotypes, and individual preferences. I call this force the *default rule*:

> If there is a default, do nothing about it.

How would that rule explain why people in the United States die because there are too few donors, whereas France has plenty? In countries such as the United States, Great Britain, and Germany, the legal default is that nobody is a donor without registering to be one. You need to opt in. In countries such as France, Austria, and

Hungary, everyone is a potential donor unless they opt out. The majority of Americans, British, French, Germans, and other nationals seem to employ the same default rule. Their behavior is a consequence of both this rule *and* the legal environment, leading to the striking contrasts between countries. Interestingly, among those who do not follow the default, most opt in but few opt out—28 percent of Americans opted in and 0.1 percent of the French opted out. If people were guided by stable preferences rather than rules of thumb, the striking differences in Figure 10-1 should not exist. In this classical economic view, the default would have little effect because people would immediately override any default that challenges their preference. After all, one only needs to sign a form to opt in, or to opt out. But the evidence indicates that it is the default rule rather than a stable preference that drives most people's behavior.

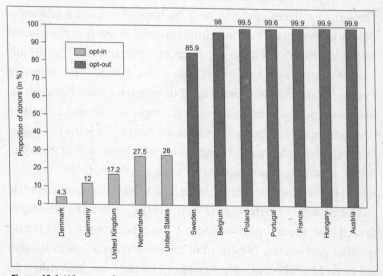

Figure 10-1: Why are so few Americans willing to donate organs? The proportion of citizens who are potential organ donors varies strikingly between countries with opt-in policies and opt-out policies. In the United States, the policy varies from state to state; some have an opt-in policy, whereas others force citizens to make a choice (based on Johnson and Goldstein, 2003).

An online experiment independently demonstrated that people tend to follow the default rule.[4] Americans were asked to assume they had just moved into a new state where the default was to be an organ donor and were given the choice to confirm or change this status. Another group was asked the same question, but the status quo was not to be a donor; a third group was required to make a choice without a default. Even in this hypothetical situation, in which sticking with the default took exactly as much effort as departing from it, the default made a difference. When people had to opt out, more than 80 percent were happy with their status as donors—a slightly higher proportion than arose with the no-default vote. Yet when people had to opt in, only half as many said they would change their status to become donors.

One possible rationale behind the default rule could be that the existing default is seen as a reasonable recommendation—primarily because it has been implemented in the first place—and following it relieves a person from many decisions. The default rule is not restricted to moral issues. For instance, the states of Pennsylvania and New Jersey offer drivers the choice between an insurance policy with an unrestricted right to sue and a cheaper one with suit restrictions.[5] The unrestricted policy is the default in Pennsylvania, whereas the restricted one is the default in New Jersey. If drivers had preferences concerning the right to sue, one would expect them to ignore the default setting, leaving little variation between the neighboring states. If they instead followed the default rule, more drivers would buy the expensive policy in Pennsylvania. And indeed, 79 percent of the Pennsylvania drivers bought full coverage, whereas only 30 percent of the New Jersey drivers did the same. It was estimated that Pennsylvania drivers spend $450 million each year on full coverage they would not have spent if the default were the same as in New Jersey, and vice versa. Thus, defaults set by institutions can have considerable impact on economic as well as moral behavior. Many people would rather avoid making an active decision, even if it means life or death.

UNDERSTANDING MORAL BEHAVIOR

My analysis of moral behavior looks at how the world *is*, rather than how it *should* be. The latter is the domain of moral philosophy. The study of moral intuitions will never replace the need for moral prudence and individual responsibility, but it can help us to understand which environments influence moral behavior and so find ways of making changes for the better.

My thesis is that humans have an innate capacity for morals just as they do for language. Children are as genetically prepared to pick up local moral rules as they are the grammar of their native language. From subculture to subculture, they learn subtle distinctions, which resemble the intricacies of local dialects, about how to behave in particular situations. In the same way that native speakers can tell a correct sentence from an incorrect one without being able to explain why, the set of rules underlying the "moral grammar" is typically not in awareness. Moral grammar, I argue, can be described by rules of thumb. Unlike in language, however, these rules are often in conflict with each other, and the result can be either morally repulsive, as in mass killing, or admirable, as with organ donation or risking one's life to save another. The underlying rule is not good or bad per se. But it can be applied to the wrong situation. I'd summarize my thoughts on moral intuitions into three principles:

- *Lack of awareness.* A moral intuition, like other gut feelings, appears quickly in consciousness, is strong enough to act upon, and its underlying rationale cannot be verbalized.

- *Roots and rules.* The intuition is attached to one of three "roots" (individual, extended family, or community) and to an emotional goal (e.g., prevent harm) and can be described by rules of thumb. These are not necessarily specific to moral behavior, but underlie other actions.
- *Social environment.* Moral behavior is contingent on the social environment. Some moral disasters can be prevented if one knows the rules guiding people's behavior and the environments triggering these rules.

Moral feelings differ with respect to the roots they are attached to: the individual, the family, or the community. A "classical" liberal, for example, understands morality to be about protecting the rights and liberties of individuals. As long as the rights of each individual are protected, people can do what they want. Other behavior is consequently not seen as a moral issue, but as the result of social conventions or a matter of personal choice. According to this individual-centered view, pornography and drug use are matters of personal taste, whereas homicide and rape are in the moral domain. Yet in other views or cultures, moral feelings extend to the family rather than to the individual alone. In a family-centered culture, each member has a role to play, such as mother, wife, and eldest son, and a lifelong obligation to the entire family. Finally, moral feelings can extend to a community of people who are related symbolically rather than genetically, by religion, local origin, or party membership. The ethics of community include principles that liberals would not acknowledge as the most important moral values, including loyalty to one's group and respect for authority. Most conservatives embrace the ethics of community and oppose what they see as the narrow moral of individual freedom. Political and religious liberals may have a hard time understanding what conservatives are talking about when they refer to "moral values" or why conservatives would want to restrict the rights of homosexuals who aren't curbing the rights of others.

The psychologist Jon Haidt proposed five evolved capacities, each like a taste bud: a sensitivity to *harm, reciprocity, hierarchy, in-group, and purity*.[6] He suggests the mind is prepared to attach moral sentiments to all or several of these, depending on the culture in which it develops. Let me connect the taste buds with the three roots. In a society with an individualistic ethic, only the first two buds are activated: to protect people from harm, and to uphold individual rights by insisting on fairness and reciprocity. According to this ethic, the right to abortion or to free speech and the rejection of torture are moral issues. Western moral psychology has been imprinted with this focus on the individual, so that from its perspective, moral feelings are about personal autonomy.

In a society with a family-oriented ethic, moral feelings concerning harm and reciprocity are rooted in the family, not in the individual. It is the welfare and honor of the family that needs protection. When it leads to nepotism, this ethic may appear suspect from the individualist point of view. In many traditional societies, however, nepotism is a moral obligation, not a crime, and smaller dynasties exist in modern democracies as well, from India to the United States. Yet while individualist societies frown on nepotism, their behavior toward family members can be in turn resented by other societies. When I first visited Russia in 1980, I found myself in a heated discussion with students who were morally outraged that we Westerners dispose of our parents when they are old, delivering them to homes where they eventually die. They found our unwillingness to take care of our own parents repulsive. A family ethic also activates a sensitivity for hierarchy. It creates emotions of respect, duty, and obedience.

In a society with a community orientation, concerns about

harm, reciprocity, and hierarchy relate to the community as its root, rather than to the family or individual. Its ethical view activates all five sensitivities, including those for ingroup and purity. Most tribes, religious groups, or nations advocate virtues of patriotism, loyalty, and heroism, and individuals from time immemorial have sacrificed their lives for their ingroup. In times of war, "support our troops" is the prevailing patriotic feeling, and criticizing them is seen as betrayal. Similarly, most communities have a code of purity, pollution, and divinity. People feel disgusted when this code is violated, be it in connection with eating dogs, sex with goats, or simply not taking a shower every day. Whereas in Western countries moral issues tend to center on personal freedom (such as the right to end one's life), in other societies, moral behavior is more focused on the ethics of community, including duty, respect, and obedience to authority, and on the ethics of divinity, such as attaining purity and sanctity.

Note that these are orientations rather than clear-cut categories. Each human society draws its moral feelings from the three roots, albeit with different emphases. The Ten Commandments of the Bible, the 613 mitzvot, or laws, of the Torah, and most other religious texts address all three. For instance, "You shall not bear false witness against your neighbor" protects the individual rights of others, "Honor your father and mother" ensures respect of familial authority, and "You shall have no other gods besides me" necessitates obeying the laws of divinity in the community. Because moral feelings are anchored in different roots, conflicts will be the rule rather than the exception.

In contrast to my view, moral psychology—like much of moral philosophy—links moral behavior with verbal reasoning and rationality. Lawrence Kohlberg's theory of cognitive development, for instance, assumes a logical progression of three levels of moral understanding (each subdivided into two stages). At the lowest level, young children define the meaning of what is right in terms of "I like it," that is, a selfish evaluation of what brings rewards and avoids punishment. At the intermediate "conventional" level, older children and adults judge what is virtuous by whether "the group approves," that is, by authority or one's reference group. At the highest "postconventional" level, what is right is defined by objective, abstract, and universal principles detached from the self or the group. In Kohlberg's words: "We claim that there is a universally valid form of rational moral thought process which all persons could articulate."[7]

The evidence for these stages comes from children's answers to verbally presented moral dilemmas, rather than from observations of actual behavior. Kohlberg's emphasis on verbalization contrasts with our first principle, lack of awareness. The ability to describe the grammatical rules of one's native language would be a poor measure of one's intuitive knowledge of the grammar. Similarly, children may have a much richer moral system than they can tell. Kohlberg's emphasis on individual rights, justice, fairness, and the welfare of people also assumes the individual to be the root of moral thinking, rather than the community or family. However, years of experimental studies do not suggest that moral growth resembles strict stages. Recall that Kohlberg's scheme has three levels, each divided in two stages; thus in theory, there are six stages. Yet stages one, five, and six rarely occur in their pure form in either children or adults; the typical child mixes stages two and three, and adults mix the two stages at the conventional level. On a worldwide scale, only 1 or 2 percent of adults were classified to be at the highest level.

I do not doubt that deliberate thinking about good and bad happens, although it may often take place after the fact to justify our actions. But here I'd like to focus on the moral behavior based on gut feelings.

I DON'T KNOW WHY, BUT I KNOW IT'S WRONG!

My first principle of moral intuitions states that people are often unaware of the reasons for their moral actions. In these cases, deliberate reasoning is the justification for, rather than the cause of, moral decisions. Consider this story:

Julie and Mark are sister and brother, traveling together in France on a summer vacation from college. One night in a cabin, they decide to make love, using both birth control pills and condoms, just to be sure. They both enjoyed making love but decided not to do it again. They kept that night a secret, which makes them feel even closer to each other. What do you think about that? Was it OK for them to make love?

Most people who hear this story feel immediately that it was wrong for the siblings to make love.[8] Only after being asked why they disapprove or even feel disgusted, however, do they begin to search for reasons. One might point out the danger of inbreeding, only to be reminded that Julie and Mark used two forms of birth control. Another begins to stutter, mumbles, and eventually exclaims, "I don't know why, but I know it's wrong!" Haidt called this state of mind "morally dumbfounded." Many of us find incest between siblings, or even cousins, repulsive, although it did not seem to have bothered the royal families of ancient Egypt. Similarly, most of us would refuse to eat the brains of our parents when they die, whereas in other cultures not doing so, leaving them to be eaten by worms, would be an insult to the deceased. According to a long philosophical tradition, the absolute truth of ethical issues can be seen intuitively, without having to reason.[9] I agree that moral intuitions often seem self-evident, but not that they are necessarily universal truths. Reasoning rarely engenders moral judgment; rather it searches to explain or justify an intuition after the fact.[10]

The second principle says that the same rules of thumb can underlie both moral actions and behavior that is not morally colored. As described above, the default rule can solve both problems that we call moral and those we do not. Another example is imitation, which guides behavior in a wide range of situations:[11]

> Do what the majority of your peers do.

This simple rule guides behavior through various states of development, from middle childhood through to teenage and adult life. It virtually guarantees social acceptance in one's peer group and conformity with the ethics of the community. Violating it may mean being called a coward or an oddball. It can steer moral action, both good and bad (donating to a charity, discriminating against minorities), as well as consumer behavior (what clothes to wear, what CDs to buy). Teenagers tend to buy Nike shoes because their peers do, and skinheads hate foreigners for no other reason than that their peers hate them.

Consider now the don't-break-ranks rule. This rule has the potential to turn a soldier both into a loyal comrade and a killer. As an American rifleman recalls about comradeship during World War II: "The reason you storm the beaches is not patriotism or bravery. It's that sense of not wanting to fail your buddies. There's sort of a special sense of kinship."[12] What appears as inconsistent behavior—how can such a nice guy act so badly; how can that nasty person suddenly be so nice?—can result from the same underlying rule. The rule itself is not good or bad per se, yet it produces actions we might applaud or condemn.

Many psychologists oppose feelings to reasons. Yet I have

argued that gut feelings themselves have a rationale based on reasons. The difference between intuition and moral deliberation is that the reasons underlying moral intuitions are typically unconscious. Thus, the relevant distinction is not between feelings and reasons, but between feelings based on unconscious reasons and deliberate reasoning.

The third principle is very practical, saying that when one knows both the mechanisms underlying moral behavior and the environments that trigger them, one can prevent or reduce moral disasters. Consider the case of organ donation. A legal system aware of the fact that rules of thumb guide behavior can make the desired option the default. In the United States, simply switching the default would save the lives of many patients who wait in vain for a donor. Setting proper defaults is a simple solution for what looks like a complex problem. Similarly, consider once again the men of the Reserve Police Battalion 101. These men grew up with the Judeo-Christian commandment "Don't murder." With his offer, Major Trapp brought this commandment into conflict with the rule "Don't break ranks." Yet Trapp could have framed his offer so that obeying the commandment wouldn't have conflicted with the need to maintain ranks. If he had asked those men who *felt up to the task* to step out, the number of men who participated in the killing would likely have been considerably smaller. Since we can't turn back the clock this is impossible to test, but both of these examples demonstrate that insight into moral intuition can influence moral behavior "from the outside."

To continue this thought experiment, imagine now the opposite: that the behavior of the reserve policemen was caused by traits such as authoritarianism, attitudes such as anti-Semitism and prejudices against minorities, or other evil motives. In these cases no such potential for immediate intervention would be possible. The social environment—Major Trapp and the other men—should make little difference, and a single policeman isolated from his comrades would have "decided" to kill, just as he did in the real situation when together with his comrades. Traits, in contrast to rules of thumb, give us little hope for change.

Moral gut feelings are based on evolved capacities. One relevant capacity is the intense identification with one's peer group that is the basis of much that makes humans unique, including the development of culture, art, and cooperation, but is also the starting point of much suffering, from social pressure for group conformity to hatred and violence against other groups. My analysis may be provocative for those who believe that moral actions are generally based on fixed preferences or independent reasoned reflection. But what may seem disillusioning in fact provides a key to avoiding moral disasters.

MORAL INSTITUTIONS

People tend to organize themselves in various forms of moral institutions, from a local neighborhood church to the Vatican, from shelters for abused women to Amnesty International. A moral institution has a code of honor or purity, defines what is decent or disgusting, and last but not least, tries to have a positive impact on society. The structure of these institutions affects the moral behavior of those who serve them, as well as the rationalization behind a member's behavior.

Bailing and Jailing

One of the initial decisions the legal system makes is whether to release a defendant on bail unconditionally or punish him with curfew or imprisonment. In the English system, magistrates, most of whom are members of the local community without legal training, are often responsible for making this decision. In England

and Wales, magistrates deal with some two million defendants every year. The work involves sitting in court for a morning or afternoon every one or two weeks, and making decisions in a bench of two or three. How should magistrates decide? The law says that they should pay regard to the nature and seriousness of the offense; to the character, community ties, and bail record of the defendant; as well as to the strength of the prosecution's case, the likely sentence if convicted, and any other factor that appears to be relevant.[13] Yet the law is silent on how magistrates should combine these pieces of information, and the legal system does not provide feedback on whether their decisions were in fact appropriate or not. The magistrates are left to their own intuitions.

How do magistrates actually make these millions of decisions? Magistrates tend to say, with confidence, that they thoroughly examine all the evidence in order to treat individuals fairly and without bias. For instance, one explained that the decision "depends on an enormous weight of balancing information, together with our experience and training." The chairman of the council stated, "We are trained to question and to assess carefully the evidence we are given."[14] As one explained self-assuredly, "You can't study magistrates' complex decision making."

The truth is that one can; people tend to believe they solve complex problems with complex strategies even if they rely on simple ones. To find out what rationale actually underlies magistrates' intuitive decisions, researchers observed several hundred hearings in two London courts over a four-month period.[15] The average time a bench spent with each case was less than ten minutes. The information available to the London magistrates included the defendants' age, race, gender, strength of community ties, seriousness of offense, kind of offense, number of offenses, relation to the victim, plea (guilty, not guilty, no plea), previous convictions, bail record, the strength of the prosecution's case, maximum penalty if convicted, circumstances of adjournment, length of adjournment, number of previous adjournments, prosecution request, defense request, previous court bail decisions, and police bail decision. In addition, they saw whether the defendant was present at the bail hearing, whether or not legally represented, and by whom. Not all of this information was present in every case, while additional information was provided in others.

Recall that the magistrates explained—and no doubt also believed—that they carefully examine all the evidence. However, an analysis of the actual bail decisions in court A revealed a simple rule that had the structure of a fast and frugal tree (Figure 10-2, left). It predicted 92 percent of all decisions correctly. When the prosecution opposed bail or requested conditional bail, the magistrates also opposed bail. If not, or if no information was available, a second reason came into play. If a previous court had imposed conditions or remand in custody, the magistrates decided the same. Otherwise they considered a third reason and based their decisions on the actions of the police. The magistrates in court B used a rule of thumb with the same structure and two of the same reasons (Figure 10-2, right).

The rules of thumb in both London courts appear to violate due process. Each bench based their decision on only one reason, such as whether the police had imposed conditions or imprisonment. One could argue that the police, or prosecution, has already looked at all the evidence concerning the defendant, and therefore magistrates simply use a shortcut—although this argument would of course make magistrates dispensable. However, the reasons in the simple tree were related neither to the nature and seriousness of the offense nor to other pieces of information relevant for due process. Furthermore, magistrates actually

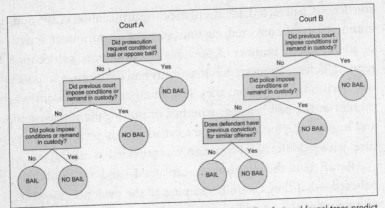

Figure 10-2: How do English magistrates make bail decisions? Two fast and frugal trees predict the majority of all decisions in two London courts. Magistrates are apparently not aware of their simple rules of thumb (based on Dhami, 2003). No bail = remand in custody or conditional bail; bail = unconditional release.

asked for information concerning the defendant, which they subsequently ignored in their decisions.[16] Unless they deliberately deceive the public (and I have no grounds to assume so), these magistrates must be largely unaware of how they make bail decisions.

Higher awareness, however, could open a moral conflict, given the ideal of due process. The magistrates' official task is to do justice both to a defendant and to the public, so they must try to avoid two errors—similar to those doctors fear—misses and false alarms. A miss occurs when a suspect is released on bail and subsequently commits another crime, threatens a witness, or does not appear in court. A false alarm occurs when a suspect is imprisoned who would not have committed any of these offenses. But a magistrate can hardly solve this task. For one, the English institutions collect no systematic information about the quality of magistrates' decisions. Even if statistics were kept about when and how often misses occur, it would be impossible to do the same for false alarms: one cannot find out whether an imprisoned person would have committed a crime had he or she been bailed. That is, the magistrates operate in an institution that does not provide feedback about how to protect the defendant and the public. Since they cannot learn how to solve the task they are meant to, they seem to try to solve a different one: to protect themselves rather than the defendant. Magistrates can only be proved to have made a bad decision if a suspect who was released fails to appear in court or commits an offense or crime while on bail. If this happens, magistrates are able to protect themselves against accusations by the media or the victims. The magistrates in court A, for instance, can always argue that neither the prosecution, nor a previous court, nor the police had imposed or requested a punitive decision. Thus, the event was not foreseeable. This defensive decision making is known as "passing the buck."

The English bail system thus requests that magistrates follow due process, but it doesn't provide the institutional setting to achieve this goal. The result is a gap between what magistrates do and what they believe they are doing. If magistrates were fully aware of what they are doing, they would come into conflict with the ideal of due process. Here is the starting point to eradicating false self-perceptions and creating the conditions to improve the English bail system.

Split-Brain Institutions

How do institutions shape moral behavior? Like the ant's behavior on the beach, human behavior adapts to the natural or social environment. Consider another institution that, like the English magistracy, requires its employees to perform a moral duty. The employee can commit two kinds of errors: false alarms and misses. If the institution does not provide systematic feedback concerning false alarms and misses, but blames the employees when a miss occurs, it fosters employees' instinct for self-protection over their desire to protect their clients and supports self-deception. I call this environmental structure a *split-brain institution*. The term is borrowed from the fascinating studies of people whose corpus callosum—the connection between the right and left cerebral hemispheres—has been severed.[17] A patient with this condition was flashed the picture of a naked body in her left visual field and began to laugh. The experimenter asked her why she was laughing, and she blamed it on his funny tie. The picture only went to her right (nonverbal) side of the brain. Because the brain was split, the left (verbal side) had to do the explaining without any information. Split-brain patients confabulate fascinating post-hoc stories with the left sides of their brains to rationalize behavior initiated by the right side. Similar processes occur in ordinary people. Neuroscientist Mike Gazzaniga, who has studied split-brain patients, calls the verbal side of the brain the *interpreter*, which comes up with a story to account for behavior produced by unconscious intelligence. I argue that a magistrate's or any other person's "interpreter" does the same when it tries to explain a gut feeling.

The analogy only holds to a point. Unlike a split-brain patient, a split-brain institution can impose moral sanctions for confabulating and punishment for awareness of one's actions. We saw that if magistrates had been fully aware that they were "passing the buck," they would have realized that their method conflicted with due process. Medical institutions, albeit not moral institutions in the narrow sense, often have a similar split-brain structure. Many Western health systems allow patients to visit a sequence of specialized doctors but do not provide systematic feedback to the doctors concerning the efficacy of their treatments. Doctors are likely to be sued for overlooking a disease but not for overtreatment and overmedication, which promotes doctors' self-protection over the protection of their patients and supports self-deception.

Transparency

Simplicity is the ink with which effective moral systems are written. The Ten Commandments are a prime example. According to the Bible, a list of religious precepts was divinely revealed to Moses on Mount Sinai. Engraved on two tablets of stone, their number was small, matching that of human digits. The ten short statements were easy to memorize and have survived for millennia. If God had hired legal advisers on Mount Sinai, they would have complicated matters by adding dozens of further clauses and amendments in an attempt to cover as many aspects of moral life as possible. Completeness, however, does not seem to have been God's goal. God, I believe, is a satisficer, not a maximizer. He concentrates on the most important issues and ignores the rest.

How many moral rules does a society need? Are ten enough or do we need a system that has the complexity of the American tax law? This law is so comprehensive that even my tax adviser cannot understand all its details. An opaque legal system fails to generate trust and compliance among citizens. Transparency and trust are two sides of the same coin. A complex legal system promotes the interests of lobbying groups who punch innumerable loopholes into its laws. The legal expert Richard Epstein has argued that the ideal of an all-encompassing legal system is an illusion. No system of any complexity can cover more than 95 percent of legal cases; the rest must be decided by judgment. Yet these 95 percent, he argued, can be resolved with a small number of laws. In his seminal book *Simple Rules for a Complex World*, Epstein, topping Moses, proposed a system of only six laws, including the right to self-ownership and protection against aggression.

A HEDONIC CALCULUS?

So far I have dealt with the question of how behavior is, rather than how it should be. In many situations, people's moral feelings are based on unconscious rules of thumb. I would not exclude deliberate reasoning as a motivation for moral behavior, but I think it occurs only in unusual contexts, such as in professional debates or in the midst of societal upheaval. Interestingly, the same distinction between simple rules and complex reasoning also exists in moral philosophy, which tries to answer the question of how people ought to behave.

The Ten Commandments exemplify the rules-of-thumb approach. The advantage of a small number of short statements such as "Honor your father and mother" and "Do not commit murder" is that they can easily be understood, memorized, and followed. Simple rules differ from what moral philosophy calls *consequentialism*,[18] in which the ends justify the means. Is it right to torture a suspected terrorist if torture might protect the safety of a country? There are two views. One argument is to consider the consequences of both alternatives (torture or no torture), their probabilities, and choose the one with the highest expected benefit. If the negative consequences of torture are small in comparison to its benefits to a country's safety, the decision is to torture. The other argument is that there are moral principles such as "Do not torture" that have absolute precedence over any other concerns.

The ideal of maximizing expected utility or happiness is the lifeblood of much moral and legal philosophy. The seventeenth-century French mathematician Blaise Pascal proposed maximizing as the answer to moral problems, such as whether or not one should believe in God.[19] He argued that this decision should not be based on blind faith or blind atheism, but on considering the consequences of each action. If one believes in God, but he does not exist, one will forgo a few worldly pleasures. However, if one does not believe in God but he exists, eternal damnation and suffering will result. Therefore, Pascal argued, however small the probability that God exists, the known consequences dictate that believing in God is rational. What counts are the consequences of actions, not the actions themselves. This way of thinking also exists in a collective rather than an individual form, best known by the maxim:

Seek the greatest happiness of the greatest number.

The English legal and social reformer Jeremy Bentham (1748–1832) proposed this guideline and supplied a calculus for actually determining the action that produces the greatest happiness.[20] His *hedonic calculus* is the felicific equivalent of Franklin's *moral algebra*, which we met in chapter 1.

The Hedonic Calculus

The value of each pleasure or pain arises from six elements, its

1. intensity,
2. duration,
3. degree of certainty,
4. remoteness,
5. fecundity (the probability of being followed by sensations of the same kind), and
6. purity (the probability of not being followed by sensations of the opposite kind).

To determine the action that likely produces the greatest happiness, and hence moral rightness, Bentham provided the following sequence of instructions for each action. Start with one person whose interests will be affected by the action. Sum up all the values for all potential pleasures and pains the person might experience, and determine the balance for the action. Repeat the process for every other person whose interests are concerned, and determine the balance for all persons. Then repeat the entire process for the next action, and finally choose the action with the highest score.

Bentham's calculus is the prototype of modern consequentialism. How would it work in our world? Assume a Boeing 747 passenger plane packed with four hundred passengers is heading toward Los Angeles on a cloudy evening. The communication between ground and cockpit suddenly breaks down, and a passenger sends a friend a text message that says the plane has been hijacked. Then there is silence. The ground crew suspect that the plane might be headed straight at the Library Tower, like the planned attack the Bush administration is reported to have foiled once. The Boeing would reach the Library Tower in five minutes, and an F-15 fighter aircraft is in the air, ready to strike. The aircraft would have to act fast to prevent the plane from descending on the target and its parts from falling into a highly populated area. At the same time, one cannot say with certainty whether an attack on the tower will happen. Would you order the F-15 pilot to shoot down the Boeing, killing four hundred innocent passengers plus crew, or not?

This scenario is both simple and complicated for the hedonic calculus. It is simple because there are only two possible actions, to shoot the plane down or to wait and see what happens. Yet it is complicated because the decision has to be made under limited time and knowledge. How many people are in the Library Tower? Is the event really a repeat of the 9/11 attacks, or was the text message an error, perhaps even a bad joke? Might the F-15 pilot shoot down the wrong plane in the cloudy sky? This situation may not be a fair example for the hedonic calculus, since the calculations of pleasures and pains involve lots of guesswork and possibilities for error. To follow the calculus, one would try to estimate for every person whose interests are concerned—each passenger, crew member, person in the tower, person on the ground nearby, and the relatives and close friends of all of these—the intensity, duration, and other dimensions of each potential pain and pleasure caused if the plane were shot down, and the same if it were not.

Although Bentham's calculus gave birth to the type of moral system that promoted democratic and liberal reforms, it is silent on the practical issues of real-time decision making. Its problem is twofold. First, if there is no known way to obtain reliable estimates of the values involved, one can select those that either advise shooting or not, justifying a decision made on other grounds. Nor is this problem limited to decisions under time constraints. The philosopher Daniel Dennett posed the question of whether the meltdown at Three Mile Island was a good thing or a bad thing.[21] In planning an action where such a meltdown could happen with some probability, should one assign a positive or negative utility? Do its long-term effects on nuclear policy, considered positive by many, outweigh its negative consequences? Many years after the event, Dennett concludes that it is still too early to say, and also too early to know when the answer will be available. Second, the advantage of complex calculations of this kind is not proved. We have already seen that even if it's possible to weigh all reasons, the result is often less accurate than that obtained from one good reason.

After the events of September 11, 2001, the plane scenario seems likely enough that countries have made rulings to cope with it. In February 2006, Germany's Federal Constitutional Court ruled that sacrificing and deliberately killing innocent citizens because of a suspected terrorist action violates the federal constitution that explicitly protects human dignity. That is, it is illegal to shoot down a hijacked plane with innocent passengers in it. The court also mentioned the danger of false alarms, that is, the possibility of shooting down a plane unnecessarily in moments of confusion and uncertainty. The Russian parliament on the other hand passed a law that does allow passenger planes suspected of being used as flying bombs to be shot down. These disparate legal decisions illustrate the conflict between consequentialism and a type of Kantian first-principles ethic that follows the rule "Don't kill innocent people as a means to an end."

These two systems differ in whether they are willing to make trade-offs. The idea that one ought to make trade-offs in order to be morally responsible often conflicts with people's gut feelings.

ARE TRADE-OFFS IMMORAL?

Diana and David, a young married couple very much in love, have started their respective careers, she as a real estate broker, he as an architect. They find the perfect spot to build their dream house and take out a mortgage. When the recession hits, they stand to lose everything they own, so they head to Las Vegas for a shot at winning the money they need. After losing at the tables, they are approached by a billionaire instantly attracted to Diana. He offers them one million dollars for a night with her.

If you and your spouse were in this financial crisis, would you accept the proposal? This, the plot of Adrian Lyne's movie *Indecent Proposal,* grapples with the morality of trade-offs. Are faithfulness, true love, and honor up for sale? Many people believe nothing would justify trading off these sacred values for money or other secular goods. Economists, however, remind us that we live in a world with scarce resources where eventually everything has its price tag, whether we like it or not. In response, Oscar Wilde is reported to have defined a cynic as someone who knows the price of everything and the value of nothing. The tension in *Indecent Proposal* arises from the conflict between treating faithfulness as a sacred value or as a commodity. The couple finally accepts the proposal, but after the night is over, they learn that their decision has exacted another price; it threatens to destroy their relationship.

Cultures differ in what they are or aren't willing to sell. So do liberal Democrats and conservative Republicans. Should the free market be extended to buying and selling body organs, PhDs, and adoption rights for children? Should people have the right to sell themselves as slaves to others? Some cultures sell their children or treat adolescent girls as a commodity to be sold to a bridegroom's family. Prostitutes earn their living by selling their bodies and sexuality, and politicians are constantly accused of having sold their ideals. If something is deemed to have a moral value, then allowing it to be traded will likely evoke moral outrage. This is one of the reasons why many citizens frown upon experts who attach a monetary value to the life of a person, dependent on age, gender, or education, as in calculations of insurances or industrial safety standards. Similarly, if an automobile company publicly announced that it did not introduce a particular safety precaution for its cars because it would have cost $100 million to save one life, moral outrage would be virtually guaranteed.[22] The overwhelming gut feeling in most cultures is that the value of lives should not be expressed in dollars.

This antipathy to trade-offs suggests that moral intuitions are based on rules of thumb that rely on one-reason decision making rather than on weighing and adding consequences. Again, there may be two kinds of people, moral maximizers who make tradeoffs and moral satisficers who don't. Most likely, every one of us has moral values that we might be willing to trade off and those that we wouldn't. The dividing line will depend on where our moral feelings have their roots. If they are rooted in the autonomy of the individual, trade-offs are unproblematic unless they do harm to other individuals and violate their rights. Yet if the moral domain is rooted in the family or community, then issues that concern hierarchy, ingroup, and purity are not up for sale.

Always laugh when people start to laugh, even though you don't understand why. The quicker, the better.

—A Japanese undergraduate at Princeton

11 : **SOCIAL INSTINCTS**

A friend told me the story of an American professor who wears the shortest skirts of any fifty-five-year-old woman he knows. While on a trip to Paris, she, a practicing Catholic, visited churches and attended masses. In one large church, visitors were separated from the pews in which people were participating in the service. Arranging to meet up with her friends later, she went to attend the service. She was the last person in a long line to take Communion, and when she finally made it into the first row, she saw the corpse of a man laid out in a coffin. All of those in front of her kissed his hands. When it was her turn, she nervously made the sign of the cross and stepped back. She then noticed that the black-clad widow and the other women were staring at her. Since she understood some French, she overheard the widow lamenting that she had always believed that her husband had no mistress and had never asked questions when he came home late. And now this! The men on the other side of the church were chuckling and nudging each other, admiring her short skirt and thinking the same thing. After all, mistresses go last in line to pay respect to the deceased. The poor professor did not know what to do; given her inadequate French and lack of time, she didn't see how to explain the situation. Deeply embarrassed, she made her way out of the church.

To an alien from Mars with no social instincts, not much would seem to have happened: the professor was in the wrong location, realized her mistake, and left. Humans with autism exhibit a similar factual view of the matter. Yet the common Homo sapiens is a social animal with a faculty for drawing quick conclusions about the dynamics of social life including betrayal, trust, and reputation. Not only do we have this capacity to go beyond the information given, but we are unable *not* to use it. We cannot stop making inferences about others. This capacity has been called social intelligence or, emphasizing its manipulative potential, Machiavellian intelligence.

But what makes us socially intelligent? According to the hypothesis of social intelligence, the social environments in which humans evolved were more complex, challenging, and unpredictable than the physical environments; therefore, this complex environment created intellectual faculties of the highest order: calculating minds that "must be able to calculate the consequences of their own behavior, to calculate the likely behavior of others, and to calculate the balance of advantage and loss."[1] In this view, the better one can read the minds of others, the more social IQ points one has. A man assesses whether a woman believes that he is in love with her; she reckons what he believes that she believes his intentions are; then he gauges what she thinks that he thinks that she thinks that he thinks what she is going to do, and so on. The more, the better. This hypothesis is based on the popular assumption that complex problems always require intricate and deliberate thought. Yet, as you may by now expect, this is not necessarily so.

I think most social interaction is, rather than the product of complex calculation, the result of special gut feelings that I call social instincts.

BASIC INSTINCTS

If a party guest argues that humans are by nature selfish, this is often taken as clear-eyed realism. In fact, many second the view that we are driven by one and only one question: "What's in it for me?" Theories of selfish egoism are hard to refute; even if people sacrifice their own interests to help others, it can be easily argued that they did so merely to feel good. I would grant that we sometimes act selfishly. Yet I also think the understanding of human nature can be improved by the realization that people carry more than one driving motivation. Selfishness is actually in conflict with two basic social instincts.

Until the spread of agriculture some ten thousand years ago, humans seem to have lived in relatively small groups. It is in these small social networks that our social instincts were shaped, the two basic ones being a *family instinct* and a *(community) tribal instinct*.[2] We share the first with our primate ancestors, whereas the second is genuinely human.

Family instinct: *Take care of your kin.*
Community instinct: *Identify with a symbolic group, cooperate, and defend its members.*

If everyone were selfish, there would be no family instinct, and, in fact many animal species appear not to have one. Most reptiles as we've seen care neither for their relatives nor their offspring after birth; some even treat them like prey. In contrast, social insects such as ants have been held up as models of sharing, caring, and community-minded creatures. Why would ants forgo reproduction to help rear the queen's offspring? That question puzzled Darwin. Today's answer is the principle of kin selection, in which individual selfishness is overcome by a disposition for helping one's relatives. In this view, if you had to choose between saving your life or the lives of your two brothers, you would be indifferent, but for three brothers you would sacrifice your life and save theirs. Your brother shares half of your genes, so from your genes' point of view, the lives of two brothers are as good as yours, but those of three are better.[3]

In reality, genes don't always get their way, but aunts and uncles do tend to invest in their nieces and nephews more than in other kids, even if they complain that the spoiled brat doesn't deserve it. The monarchy is the archetype of government by family instinct, with princes and princesses being privileged by kinship rather than merit. In many traditional societies, as mentioned before, nepotism is not a crime but a familial obligation. This family instinct infects governments when politicians promote their sons, sisters, or brothers because they are kin rather than the best person for the job.

The community instinct, however, makes us different from all other animals. It enables us to identify with a larger, symbolically marked group of people, such as a tribe, a religion, or nationality. Most people long to belong to a social group beyond their family, and emotionally attach themselves to this group, be it Texans, Shriners, or Harvard alumni. Many are willing to live and die for their ethnic group or their religion. The weird fact that many men's emotional life rotates around a ball—baseball, basketball, or football—seems to spring from the same community instinct. If you get excited by watching your home team, be it the Red Sox or the Buffalo Bills, but feel little stimulation in watching the games played by other teams, even if the quality is higher, then you follow your tribal instinct. If your first pleasure is the quality of the game, rather than the success of your home team, you have freed your love of sports from your tribal identification. Few have. When the American press reports on the Olympics, they report almost solely on American athletes, even if someone else won the events, while the Italian media report on Italians, and so on. Team sports, it seems, are not about sports per se but exist to satisfy our community instinct.

Why did this community instinct evolve? Darwin proposed one answer:

> A tribe including many members who, from possessing in a high degree the spirit of patriotism, fidelity, obedience, courage, and sympathy, were always ready to aid one another, and to sacrifice themselves for the common good, would be victorious over most other tribes; and this would be natural selection.[4]

Consistent with Darwin's view, anthropological studies indicate that most traditional human cultures are tightly regulated by social norms that support loyalty and generosity toward all members of the group, and so reduce internal conflict.[5] Conformity is secured with respect and cooperation, and deviance is punished with disrespect, ridicule, and the withdrawal of cooperation. In a war, sacrificing one's life for the group without hesitation is praised as heroism. Those who deviate from this standard of conduct are censored and punished by the others, but the norms are usually so internalized that they do not require enforcement.

The community instinct, however, has not eliminated the older family instinct, and the two can sharply conflict. When a politician arranges for relatives to occupy key political positions or even creates a dynasty, his family instinct may do disservice to the country. Wartime sets up another forum for these conflicting instincts. When parents send their offspring to war, feelings of patriotism and loyalty conflict with feelings of responsibility for their children. Moral outrage can result when powerful people manage to place their family interests over their loyalty to their country, such as when the news spread that out of all U.S. senators and congressmen only one had a son fighting in Iraq.

Identification and competition are two sides of the same coin. Community instincts cannot be put to work unless there are competing tribes that are easily distinguishable from each other. Dialects and skin color are often used to define the borders between communities, but more often there are symbolic markers. These include dress codes, religious objects, and flags. Men have given their lives to defend precious religious objects from misuse or flags from being captured. It appears that any symbol can be used to define a group, even when it's arbitrarily created. The *minimal group* experiments by social psychologist Henri Tajfel demonstrated this phenomenon. Of Polish-Jewish parentage, Tajfel lost nearly all of his family and friends in the Holocaust and developed an abiding interest in how group identities are formed, how genocide can happen, and how to end the suffering of those who are in the wrong group at the wrong time: Jews in an anti-Semitic world, foreigners in a xenophobic country, or women in a sexist culture. In his experiments, he randomly divided people into groups. No matter what group a person happened to be in, he or she quickly began to discriminate in favor of the "ingroup" members and against the "outgroup" people. Yet, if asked, people were not always aware of why they did what they did. Similar to the after-the-fact justification of the split-brain patients, ingroup members justified their discriminating behavior with rational arguments about how unpleasant and immoral the outgroup people were. These experiments investigated under controlled conditions what one can observe in many schoolyards. Watch how children tend to spontaneously form groups, stand together, and treat those not in the gang.

The community instinct is based on reciprocity. In *The Descent of Man*, Darwin arrived at the conclusion that reciprocity was the foundation stone of morality. Darwin called reciprocity—what I give to you, you will return to me—a social instinct. The exchange can be of goods or money but also of moral approval and disapproval. I support your beliefs, struggles, and sacred values, and I expect that you support mine in return. Social contracts are based on the combination of trust and reciprocity. Tit for tat, for instance, is a way to interact with others in which one trusts first, and then reciprocates (see chapter 3). I trust you by providing you with something, and I expect that you reciprocate in kind. In contrast, blind trust would not work in a society over the long run because cheaters would emerge to take the benefits without paying the costs. Therefore, the human mind also has a machinery of capabilities for protecting social contracts against exploitation. One is an automatic attention device that spots situations in which one is being cheated.[6] To detect and expel these cheaters, the human mind needs capacities such as face and voice recognition, as well as emotional devices such as feelings of guilt, ridicule, anger, and punishment.

The family instinct that favors kin and the community instinct that favors identification with unrelated members of symbolic groups are two roots of moral and altruistic behavior. These basic instincts are put to work by special social abilities such as those that allow us to detect cheaters and to trust. Let's have a closer look at trust, the cement of society.

TRUST

A man's facial expression provides cues for inferring whether he is trustworthy or not. These cues were exploited in the successful 1960 Democratic election campaign against the Republican presidential candidate, Richard Nixon. It showed a picture of Nixon, thin lipped, unshaven, with dark shadows under his eyes. The caption read: "Would you buy a used car from this man?" Trust ranks high in a modern democracy. Despite fancy information technology, a flourishing insurance industry, and a jungle of laws, few economic transactions and personal relations could thrive without at least some measure of trust in the other.

One might think that trust has always been the cement of social life, as people complain that only in the good old days could they count on others. Yet as cultural historian Ute Frevert has argued, trust in fellow humans is conspicuously rare in premodern societies.[7] Martin Luther warned people against trusting each other and admonished them to trust in God. During the nineteenth century, however, while trust in God declined, social trust grew—albeit only between certain groups of people. Men trusted men, the husband his wife, family members other family members, but trust between unmarried men and women was viewed with suspicion. Various changes in the structure of work and the transition of living in relatively small towns to large cities made trust a central issue: the new large-scale division of labor that forced workers to rely on each other; larger groups of people in which it was harder to keep tabs on everyone; and increasing mobility. Primitive societies get by with less trust: in small groups, it is possible to watch each other all the time. The more you are able to control and predict the behavior of others, the less need you have for trust.

Cooperation in an uncertain technological world requires a tremendous amount of trust, making it the lifeblood of a modern community instinct. We entrust our money to banks, open the door when the doorbell rings, and give our credit card number to strangers on the phone. If we find our home robbed by someone, we feel angry, but when the robber is our baby-sitter, we feel both angry and betrayed. The injury inflicted by the baby-sitter, destroying trust, is both material and psychological. Without trust, there would be no lasting large-scale cooperation between people, little trade, and few happy couples. Why is that? Benjamin Franklin once said, "In this world there is nothing certain but death and taxes." Social uncertainty in large societies is the problem that trust can and does help to resolve.

TRANSPARENCY CREATES TRUST

"If I seem unduly clear to you, you must have misunderstood what I said," remarked Alan Greenspan, the former chairman of the Federal Reserve Board, to a congressman. Another of his famous rejoinders was "I know you believe you understand what you think I said, but I am not sure you realize that what you heard is not what I meant."[8] It is not always possible to tell whether Greenspan spoke his mind on these occasions or wanted to preserve the legend surrounding his Delphic language, known as *Greenspanese*. As much as Greenspan was admired as a "macroeconomic magician," it was clear that once he left office, nobody would be able to continue his policy—all his tacit knowledge and expert hunches seem to be buried in his mind.

Mervyn King is governor of the Bank of England and hence the British equivalent of the chairman of the Federal Reserve Board in the United States. Over lunch, he told me the following story. When he joined the Bank of England, he asked Paul Volcker, Alan Greenspan's predecessor, whether he could advise him on how to succeed in his new job. Volcker gave his advice in one word: "mystique." King, however, decided against this policy and chose the opposite approach to dealing with the public: transparency. When the Bank of England estimates next year's inflation rate, it does not just present a number, such as 1.2 percent, as if it were an undisputed fact. Rather, it posts the board discussion on the Internet, including all arguments in favor of or against an estimate, making the decision processes accessible to everyone. The Bank also makes it clear that the prediction is not certain, as a single number would suggest, and specifies the region of uncertainty, such as between .8 percent and 1.5 percent. When King introduced this transparent system, some politicians reacted in surprise: "Are you saying that you cannot predict with certainty?" The truth is that certainty is an illusion. Being open about uncertainties can help prevent crises by alerting policy makers to the problems on the horizon. Within a decade, the policy of transparency turned the Bank of England into one of the most trusted institutions in Britain. When King leaves office, everyone will know how to continue his policy. In King's words, "Transparency is not simply a question of making available certain data. It is an approach to economic policy, almost a way of life."[9]

In some countries, politicians are advised to withhold any trace of uncertainty from the public under the pretense of "protecting" their citizens as if they were children. Yet the public is intelligent enough to see through this game, and such politicians create a climate of disbelief that generates public disinterest and political apathy. A Gallup Poll surveyed citizens across forty-seven countries and found that parliaments and congresses, supposedly the key democratic bodies, were the least trusted of all institutions.[10] Even global companies and trade unions elicited more confidence.

The policies of mystique and illusionary certainty damage public trust in institutions and compliance with the law. As the case of the Bank of England shows, there is a viable alternative, transparency, that can create both trust and an informed citizenry.

IMITATION

If you have ever opened a book on decision making, you have likely run across the idea that the human mind is an ever-busy accountant of pros and cons making dozens or even hundreds of decisions a day. Wouldn't it be more realistic to ask how people can avoid making decisions all the time? No mind or machine should try to make all decisions by itself, given the limited information and time at its disposal.[11] Often it is reasonable to ask for others' advice, or not to ask at all but simply to imitate their behavior. Many Americans change their clothes once or even twice a day, whereas most Europeans wear them for several days before washing them. No matter what norm of cleanliness is followed, it is not decided upon every morning but is a habit that results from copying others. As children, we imitate what Mom and Dad eat and how they talk; later in life we follow public and professional role models. Imitation is not simply a shortcut for deliberate decisions when one has little knowledge and time, but is one of the three processes—the others being teaching and language—that allow for the vast cultural transmission of information over generations. Without these, every child would have to start from scratch in the world and learn by individual experience. Most animals live without this kind of cultural learning. Even in other primates, there is limited imitation, little teaching, and only rudimentary forms of language. I shall distinguish two basic forms of imitation:[12]

Do what the majority of your peers do.
Do what a successful person does.

If you find an eccentric person admirable and begin to imitate her extravagant ways rather than those of your more conventional friends, you are not following the majority. If, however, you find her behavior intolerable and act like your other friends, you are. This rule of thumb shapes our intuitions about what we want and dislike, respect and scorn. We are prone to join unquestioningly the screams of Rolling Stones fans or the roaring hordes of Harley-Davidson bikers if that's what our peers do. Imitation of the majority satisfies the community instinct, because belonging to a group creates comfortable conformity and a distinction from outside groups. Similarly, imitating a successful group member can enhance future status in that group, and if others do the same, also strengthens conformity.

Neither form of imitation is good or bad per se. In technological invention and industrial design, imitating the successful is a major strategy. The Wright brothers successfully relied on this copycat rule by patterning their flight machines after Octave Chanute's glider, whereas others were doomed to fail by trying instead to imitate the flight of the albatross and the bat. The success of imitation depends again on the structure of the environment. Structural features that can make imitation adaptive include

- *a relatively stable environment,*
- *lack of feedback, and*
- *dangerous consequences for mistakes.*

Imitation can pay in a stable environment. How should a son run his father's company? When the business world in which the company operates is relatively stable, the son may be well advised to imitate the successful father, rather than starting from scratch by introducing new policies with unknown consequences.

Imitation can also pay in a world with little feedback. We often cannot find out whether an action we took was actually better than the alternative we did not take. Will children become better moral beings if parents are strict or if they are allowed to do whatever they want? The answer is almost impossible to find out by experience. Most people have only a few children, and it takes a long time to see the results of their upbringing. And even then parents still do not know what the results would have been if they had acted differently. Limited feedback is typical for unique decisions, such as what to do after college, and for repeated events whose consequences can only be observed after a long time, such as lifestyle. In these cases, imitation can pay, whereas individual learning has its natural limits.

Imitation can also pay in situations with dangerous consequences. Food choice is a case in point. Relying only on individual experience to learn which berries found in the forest are poisonous is obviously a bad strategy. Here, imitation can save your life—although it may cause false alarms. When I was a boy, I was told on good authority never to drink water after eating cherries or I would get very sick or even die. Where I grew up, everyone acted this way, and so did I. Nobody asked why. One day I shared an ample serving of cherries with a British friend who had never heard of this danger. When he reached for a glass of water, I tried to stop him and save his life, but he only laughed. He took a sip, and nothing happened, curing me of that belief. I still won't reheat mushrooms, however, simply because I've been warned that doing so is dangerous.

When is imitation futile? As mentioned before, when the world is quickly changing, imitation can be inferior to individual learning. Consider once again the son who inherits his father's firm and copies his successful practices, which have made a fortune over decades. Yet when the environment changes quickly, as in the globalization of the market, the formerly winning strategy can cause bankruptcy. In general, imitating traditional practice tends to be successful when changes are slow, and futile when changes are fast.

CULTURAL CHANGE

Imitation is a fast way to acquire the skills and values of a culture and to keep cultural evolution in motion. If everyone imitated everyone, however, change would be impossible. Social change, it seems, can be the product of psychological factors as well as of economic and evolutionary processes.[13]

Not Knowing the Rules Can Change the Rules

Social change has been brought about by multiple means, including admirable acts of individual courage. In 1955, when a black woman named Rosa Parks refused to surrender her bus seat to a white man in Montgomery, Alabama, she was arrested for violating the city's racial segregation law. Black activists, led by the young Martin Luther King Jr., boycotted the transit system for more than a year, during which King's home was dynamited and his family threatened, until they achieved the goal of desegregating the bus system. Parks's decision can be said to have ignited the U.S. civil rights movement. Her courage in not following the law and her willingness to undergo punishment for an ideal provide inspiring examples of the psychological factors that promote change. Yet there are more unlikely candidates.

A dear friend of mine is a professor at a leading American university and is now on the verge of retiring after a brilliant career. After becoming a teenage beauty queen, she focused her time and attention on her studies and in the mid-1950s completed her B.A. with distinction. After getting her degree, she asked her adviser what she needed to do next to have the academic career she envisioned and whether he would write her a letter of recommendation for Harvard and Yale. The adviser looked at her with astonishment: "My dear, you are a woman! No. I will not write a recommendation. You are much too smart. You might take away a job from a man."

My friend was shocked and close to tears. She had never realized that there was a simple reason why all of her professors were male: women were not meant to be in this profession. When her adviser so adamantly rejected her request, she felt no anger about the unwritten rules of male academia but was instead deeply embarrassed by her faux pas. Overwhelmed by her emotional reaction and determination, however, her other professors decided to do what they would have done for a male student, and wrote her recommendations. Eventually she became one of the first female professors in her field. Her innocent ignorance helped to open the door for her career, and she became a role model for many women. I suspect that if she had been more conscious of her place as a female in male academia, as other women were, she might not have even tried.

The power of ignorance to speed up social change is a common plot in literature. Among the many heroes in Wagner's *Ring of the Nibelung*, Siegfried is the most clueless one. Siegfried grows up without parental care. He is the naive hero who acts impulsively, whose adventures happen to him, rather than being deliberately planned. Siegfried's combination of ignorance and fearlessness is the weapon that eventually brings down the rule of the Gods. Similar to Siegfried is Parsifal, the hero of Wagner's last work. Brought up by his mother in a lonely forest, he knows nothing about the world when he begins his quest for the Holy Grail. The strength of Siegfried, Parsifal, and similar characters lies in their not knowing the laws that rule the social world. Like my colleague when she was young, the naive hero has a seemingly reckless and childlike disregard for social conventions. Ignorance of the status quo, and so a lack of respect for it, is a great weapon with which to revolutionize the social order.

These heroes' intuitive actions were based on missing knowledge, but the gut feeling "I can do it!" can succeed even when based on false information. Christopher Columbus had problems financing his dream of finding a western sea route to India. His contemporaries believed he had miscalculated the distance to India, and they were right. Though Columbus knew that the earth was round, he widely underestimated its radius. Eventually Columbus got the funds, sailed off, and came across something else: America. Had he known that India was so far away, he might not have even set sail. Note that Columbus himself did not see his discovery as serendipity; he insisted up to his death that he had made it to India.

Can one use the positive potential of ignorance systematically rather than accidentally? For instance, when I became director at the Max Planck Institute for Human Development, I "inherited" several staff members from my predecessor. Immediately, well-meaning people offered to give me the details on the staff's social and professional flaws. I declined. My policy was not to try to know everything about my employees but to give them a chance for change. Professional tensions are created not only by individual staff members but also by the environment in which they work, including those people who volunteer to complain about them. Since I was creating a new environment, these employees were given a chance to escape the picture others made of them—a chance they all took.

Ignorance can be powerful but is not a value per se. It can help to promote social change in situations like those described here, but it is far from a universal recipe. All of my stories involve substantial degrees of uncertainty or social unpredictability; ignorance would be of little help in routine day-to-day problem solving where efficiency and expertise are wanted.

Embarrassment

On the British Isle of Wight in 2003, conditions on the school bus got very bad.[14] Pupils were fighting, abusing each other, even throwing seats out of the windows and distracting the driver from paying attention to the road. The behavior of a minority of children was putting the safety of everyone else at risk. The bus drivers did not like to leave the troublemakers at the roadside, but eventually found themselves forced to do that or call the police. But even these harsh measures didn't help. The crime and disorder manager for the Isle of Wight then introduced a simple but effective measure. She separated the rowdy boys from their peers and transported them in a pink bus called the "Pink Peril." The bus and its color were carefully selected. It was the oldest vehicle the company owned, had no heating, and had been painted the color regarded as most uncool by boys, who made up most of the troublemakers. Unruly pupils were embarrassed to be seen on the bus, and either covered their faces or slouched down below the windows to prevent people from staring at them. As a result, there was a substantial reduction in violent bus incidents. Embarrassment proved to be more effective than the methods of the police.

The idea of using social emotion as a deterrent rather than physical punishment is nothing new. In medieval Europe, various offenders were forced to wear masks of shame to publicly display their misdemeanors. A flute-of-shame mask was for bad musicians, a swine mask for men treating women poorly, and a hood of shame for bad students. Designing environments that elicit offenders' emotions—embarrassment, shame, and feelings of guilt—can be a powerful way to deal with deviant behavior, whatever its causes are.

Ridicule is another effective instrument for influencing people's behavior and beliefs. In the Bavarian town where my grandfather lived, people often had difficulties in sleeping and woke up from nightmares, which made them dread falling asleep again. The cause of this widespread suffering was common knowledge: a witchlike being, with hairy hands and feet, called the Trud. In the night, while you slept, she would sit on your chest so heavily that you could barely breathe. She particularly liked to plague pregnant women and deer. In Bavaria and Austria there was an elaborate folklore of stories about men and women suffering the torments of the Trud, some even being killed by suffocation. As with other folklore, it was hard to refute the reality of the Trud, since so many adults had encountered her. Rational arguments against her existence proved ineffective or were thrust aside.

All that changed during World War II, when soldiers were put up in towns all across Bavaria. They took quarter in farmhouses and shared meals with the host family and their servants. Over dinner, some farmer complained about waking up in the night out of breath because the Trud was, once again, sitting on his chest. The soldiers at the table had never heard of the Trud before and began to chuckle. The locals insisted on the truth of their story, but the soldiers responded with bursts of laughter. After a few outbursts, the embarrassed locals stopped talking about the Trud for fear of being ridiculed. The silent farmers may have continued to believe in her existence, but the fact that they no longer dared to tell the stories in public erased the Trud from the collective memory of the following generations. Today, few in Bavaria have heard of the Trud, and nightmares are attributed to other causes.

Rumor Tears Down the Wall

On one mild November night in 1989, the Berlin Wall fell. Shortly before midnight on November 9, thousands of East Germans pushed their way through the first checkpoint, and by 1:00 a.m. all borders were wide open. On that memorable night, Berliners danced on the Wall in front of the Brandenburg Gate and cheered on an endless stream of East Berliners going West. People got together with flowers in their hands and tears in their eyes. No

Figure 11-1: No one predicted the fall of the Berlin Wall. © ullstein bild-fishan.

politician had predicted the fall of the Wall, even one day beforehand. "Who got us into this mess?" asked the perplexed East German prime minister in dismay. "That's impossible. It's incredible!" rejoiced the West German chancellor.[15] Everyone, including the CIA and President George H. W. Bush, was taken by total surprise.

The Wall had divided Berlin for almost thirty years. Rising up to fifteen feet high, it extended for twenty-eight miles through the city, made of concrete, topped with barbed wire, and guarded by watchtowers, mines, and a special police force. More than one hundred East Germans had been killed trying to cross it and escape into the West, and thousands were captured in the attempt. In early 1989, the East German prime minister had announced that the Wall would still stand in fifty or even a hundred years.[16] There seemed little hope that East Germans would be allowed to travel freely. Yet their government came under strong pressure after a reformist Hungarian government opened its borders to the West, allowing East Germans to escape via Hungary. When Czechoslovakia also opened its borders, thousands used this shorter route in a mass exodus for West Germany. Every Monday, large crowds of East Germans demonstrated for the basic rights of citizens in a democracy: free travel, free press, and free elections. The scene was set for political change, but no one knew how to make it happen.

On November 9, the East German government reacted by announcing new guidelines for traveling outside of the country, which offered only slight relief over the old handcuffs. Citizens still had to apply first for a passport (most had none), and then for a visa. These applications used to take months of endless bureaucratic paperwork, after which the visa could still be denied. The new guidelines promised only that this process would be sped up. At 6:00 p.m., Günter Schabowski, the new secretary of the East German Central Committee for Media Politics, held an hour-long press conference, only mentioning the new guidelines at the very end. Having missed the government meeting where these had been discussed, the tired-looking and overworked Schabowski hemmed and hawed his way through an obviously unfamiliar text. Informed and attentive listeners recognized that there was not much new—East German politics as usual. An Italian journalist asked when the new regulations would be effective. Schabowski, who did not seem to know, hesitated, looked at his sheet of paper, and then said "right now, immediately." At 7:00 p.m., he ended the conference.

Whereas most reporters saw little reason for excitement, the Italian journalist rushed out and, shortly afterward, his agency spread the news that "the Wall fell." This report had no backing from what Schabowski had said. Simultaneously, an American reporter who did not understand German interpreted the translation of the conference as meaning that the Wall was now open, and NBC broadcasted that from tomorrow morning on, East Germans could traverse the Berlin Wall without any restrictions. At 8:00 p.m., the West German TV news, under time pressure, summarized the press conference in their own words, and Schabowski was shown saying "right now, immediately." At the end of the report, the headline "East Germany Opens Border" was added. Other news agencies entered this contest in wishful thinking and mistakenly reported that the border was already open. A waiter from a nearby café in West Berlin went with his guests and a tablet of champagne glasses to the perplexed border guards to make a toast to the opening of the Wall. The guards, who thought this was a bad joke, refused and sent the troublemakers back. Yet the rumor spread to the West German parliament in Bonn, which happened to be meeting at that time. Deeply moved, some with tears in their eyes, the representatives stood up and began to sing the German national anthem. The East Germans who were watching West German television were more than willing to engage in the wishful thinking seeded by the news. A dream infinitely far away seemed to have come true. Thousands and soon tens of thousands of East Berliners jumped into their cars or walked to the border crossings to the West. Yet the guards had, of course, no orders to open the border. Angry citizens demanded what they believed was their new right of way, and the guards at first refused. Yet in the face of an avalanche of citizens physically pushing at them, an officer at one crossing eventually opened the barriers, fearing that his men would otherwise be trampled to death. Soon all the crossings were open. No shot was fired, no blood spilled.

How could this miracle happen? Years of diplomatic negotiations and financial payments from the West had failed. The immediate cause for the fall of the Berlin Wall turned out to be a combination of wishful thinking and a subsequent unsubstantiated rumor that spread like wildfire. The government was as surprised as its citizens, whereas a well-planned uprising could easily have been suppressed by tanks and soldiers, as had happened in 1953. Wishful thinking could spread because the new guidelines had been put together in a hurry, without the standard press release to ensure that journalists reported what they were supposed to. Yet if the media and the citizens of Berlin had carefully listened to what Schabowski had said and looked at the facts, nothing would have happened on this remarkable night, and the next day would have been just another day in a divided Berlin. After the Wall fell, however, both the prime minister and Schabowski quickly did an about-face and took credit for the opening of the border.

Rumor and wishful thinking are virtually always seen as negative, to be avoided and replaced by deliberate, informed reasoning. Like deliberation and negotiation, however, they can be positive in a powerful way. A high-level West German government official reportedly concluded after the press conference that East German politics had once again failed to change, and went to bed. He slept through this historical night because he knew too much.

* * *

In Western thought, intuition began as the most certain form of knowledge and has ended up being scorned as a fickle and unreliable guide to life. Angels and spiritual beings were once thought to intuit with impeccable clarity—superior to merely human ratiocination—and philosophers argued that intuition made us "see" the self-evident truths in mathematics and morals. Yet today, intuition is increasingly linked to our bowels rather than our brains, and has descended from angelic certainty to mere sentiment. But gut feelings are in fact neither impeccable nor stupid. As I have argued, they take advantage of the evolved capacities of the brain and are based on rules of thumb that enable us to act fast and with astounding accuracy. The quality of intuition lies in the intelligence of the unconscious: the ability to know without thinking which rule to rely on in which situation. We have seen that gut feelings can outwit the most sophisticated reasoning and computational strategies, and we have seen how they can be exploited and lead us astray. Yet there is no way around intuition; we could achieve little without it.

In this book, I invited you into the largely unknown land of intuitive feelings, shrouded in a mist of uncertainty. For me, it has been a fascinating journey, full of surprises about the power of gut feelings and the wonders that emerge when the haze lifts. I hope that you too have enjoyed these glimpses into the intelligence of the unconscious and learned that there are many good reasons to trust your gut.

ACKNOWLEDGMENTS

Gut Feelings is inspired by the research I conducted over the past seven years at the Max Planck Institute for Human Development. This book is intended to be an entertaining and readable exposition of what we know about intuitions and is purposely not written as an academic text. Mixing real stories and psychological concepts, I hope to motivate the reader to take gut feelings more seriously and to understand where they come from. For those who get hooked on the topic, the relevant references to the scholarly literature are provided.

Many dear friends and colleagues were so kind to read, comment on, and improve this book manuscript in its various stages: Peter Ayton, Lucas Bachmann, Simon Baron-Cohen, Nathan Berg, Sian Beilock, Henry Brighton, Arndt Bröder, Helena Cronin, Uwe Czienkowski, Sebastian Czyzykowski, Lorraine Daston, Mandeep Dhami, Jeff Elman, Ursula Flitner, Wolfgang Gaissmaier, Thalia Gigerenzer, Daniel Goldstein, Lee Green, Dagmar Gülow, Jonathan Haidt, Peter Hammerstein, Ralph Hertwig, Ulrich Hoffrage, Dan Horan, John Hutchinson, Tim Johnson, Günther Jonitz, Konstantinos Katsikopoulos, Monika Keller, Mervyn King, Hartmut Kliemt, Elke Kurz-Milcke, Julian Marewski, Laura Martignon, Craig McKenzie, Daniel Merenstein, John

Monahan, Wiebke Möller, Andreas Ortmann, Thorsten Pachur, Markus Raab, Torsten Reimer, Jürgen Rossbach, Erna Schiwietz, Lael Schooler, Dennis Shaffer, Joan Silk, Paul Sniderman, Masanori Takezawa, Peter Todd, Alex Todorov, and Maren Wöll.

My special thanks go to Rona Unrau, who edited the entire book manuscript, including the footnotes and references. More than that, she assisted in researching several topics and insisted on achieving clarity in exposition. She has been a wonderful support. Hilary Redmon, Juli Barbato, and Katherine Griggs from Viking were also of enormous help in giving the book a final thorough polish. Lorraine Daston, my wife, and Thalia Gigerenzer, my daughter, provided intellectual and emotional support during the four years I was working on this book. People are crucial, as is the environment in which one works. I was lucky to have the unique support of the Max Planck Society and to profit from the outstanding resources and the splendid intellectual atmosphere. It's research paradise.

NOTES

PART 1: UNCONSCIOUS INTELLIGENCE

Chapter 1: Gut Feelings

1. Franklin, 1779. The scientist and statesman Benjamin Franklin was one of the great figures of the Enlightenment, and his moral algebra is an early version of modern utilitarianism and rational choice theory. In his ethics, the rake and the drunkard are just like other people but became that way simply because they failed to calculate correctly.

2. Wilson et al., 1993. Similarly, Halberstadt and Levine, 1999, and Wilson and Schooler, 1991, experimentally showed that introspection can reduce the quality of decisions, and Zajonc, 1980, and Wilson, 2002, provided further stories on the conflict between balance sheets and gut feelings. Wilson reports of a social psychologist who tried to decide whether to accept a job offer from another university by using a balance sheet, listing pros and cons. Halfway through, she said, "Oh hell, it's not coming out right! Have to find a way to get some pluses on the other side" (167).

3. Schwartz et al., 2002. The term *satisficing* was introduced by the Nobel laureate Herbert A. Simon and originated in Northumbria, a region in England on the Scottish border, where it means "to satisfy."

4. Goldstein and Gigerenzer, 2002. The term *heuristic* is of Greek origin and means "serving to find out or discover." The Stanford mathematician G. Polya, 1954, distinguished between heuristic and analytic thinking. For instance, heuristic thinking is indispensable for finding a

mathematical proof, whereas analytic thinking is necessary for checking the steps of a proof. Polya introduced Herbert Simon to heuristics, and it is on the latter's work that I draw. Independently, Kahneman, et al., 1982, promoted the idea that people rely on heuristics when making judgments but focused on errors in reasoning. In this book, I use *heuristic* and *rule of thumb* as synonyms. A heuristic, or rule of thumb, is fast and frugal; that is, it needs only minimal information to solve a problem.

5. Dawkins, 1989, 96.

6. Babler and Dannemiller, 1993; Saxberg, 1987; Todd, 1981.

7. McBeath et al., 1995; Shaffer et al., 2004.

8. McBeath et al., 2002; Shaffer and McBeath, 2005.

9. Sailors learn that if another boat approaches, and the bearing remains constant, a collision will occur. Soldiers are taught that if mortar or shells are fired at them, they should wait until the object is high up in the air and then point at it. If it does not move relative to their finger, they'd better run. If the object descends below their finger, it is going to land in front of them, and if it continues to ascend, it will land behind them.

10. Collett and Land, 1975; Lanchester and Mark, 1975.

11. Shaffer et al., 2004.

12. Dowd, 2003.

13. Horan, in press.

14. Lerner, 2006.

15. Commercial egg producers want to identify female chickens quickly and avoid feeding the unwanted sex: the unproductive males. Before the art of chicken sexing emerged in Japan, poultry owners had to wait until chicks were five to six weeks old; today, expert chicken sexers can reliably determine the sex of day-old chicks on the basis of very subtle cues, at a speed of some thousand chickens per hour. R. D. Martin, author of *The Specialist Chick Sexer* (1994), cites an expert on his publisher's Web site: "There was nothing there but I knew it was a cockerel" and comments, "This was intuition at work." Like other tacit skills, sexing can become an obsession. "If I went for more than four days without chick sexing work I started to have 'withdrawal symptoms.'" http://www.bernalpublishing.com/poultry/essays/essay12.shtml.

16. Yet this belief is alive and kicking. Even when it comes to emotional in-

telligence, it is still assumed that it can be measured by asking people questions concerning declarative knowledge, for instance, to rate themselves on the statement "I know why my emotions change" (see Matthews, et al., 2004). The underlying belief is that people are able and willing to tell how their intelligence functions. In contrast, the influential work by Nisbett and Wilson, 1977, reviewed experimental evidence that people often do not have introspective access to the reasons underlying their judgments and feelings. Research on *implicit learning* refers to learning that proceeds both unintentionally and unconsciously (Lieberman, 2000; Shanks, 2005).

17. For similar definitions, see Bruner, 1960, Haidt, 2001, and Simon, 1992.

18. See Jones, 1953, 327, on Freud, and Kahneman et al., 1982, on cognitive illusions. For my critique of these views, see Gigerenzer, 1996, 2000, 2001; for a reply to my critique, see Kahneman and Tversky, 1996, and Vranas, 2001.

19. For instance, Gladwell's (2005) engaging book *Blink* features research, including my own, on how people make successful snap judgments: "and—*blink!*—he just knows. But there's the catch: much to Braden's frustration, he simply cannot figure out *how* he knows" (49). In this book, I try to explain how these intuitions work.

20. Wilson, et al., 1993, 332, explained why women who gave reasons were less satisfied with the posters they chose in this way: "Introspection . . . can change an optimal weighing scheme into a suboptimal one. When people analyze reasons, they might focus on those attributes of the attitude object that seem like plausible causes of the evaluations but were not weighted heavily before." The idea that the process underlying intuitive judgments resembles Franklin's balance sheet and rational choice theory is also found in Dijksterhuis and Nordgren, 2006, and Levine et al., 1996. These distinguished researchers demonstrate in their fascinating experiments that less thought can be more. To explain this phenomenon, however, they do not push the vision that less could actually be more (see next chapters) but instead assume that instantaneous judgments, if they are good, must be based on unconscious calculations of all pros and cons.

21. This includes Ambady and Rosenthal, 1993, Cosmides and Tooby, 1992, Gazzaniga, 1998, Hogarth, 2001, Kahneman et al., 1982, Myers, 2002, Payne et al., 1993, Pinker, 1997, and Wegner, 2002. For an introduction

to the research at the Max Planck Institute see Gigerenzer et al., 1999, Gigerenzer and Selten, 2001, Gigerenzer, 2004a, and Todd and Gigerenzer, 2003.

Chapter 2: Less Is (Sometimes) More

1. Einstein is quoted in Malkiel, 1985, 210. For Einstein, the simplicity of an explanation was both a sign for its truth and a goal of science: "Our experience up to date justifies us in feeling sure that in Nature is actualized the ideal of mathematical simplicity" (Einstein, 1933, 12).

2. Bursztajn et al., 1990.

3. Luria, 1968, 64.

4. Anderson and Schooler, 2000; Schacter, 2001; Schooler and Hertwig, 2005.

5. James, 1890/1981, 680. The Word buffer analogy to memory retrieval in Figure 2-1 was proposed by Lael Schooler and is based on Schooler and Anderson, 1997.

6. That is, it was unable to pick up concepts such as noun-verb agreement in embedded clauses (Elman, 1993; see also Newport, 1990).

7. Cited in Clark, 1971, 10.

8. Huberman and Jiang, 2006.

9. DeMiguel et al., 2006. The optimal asset allocation policies included sample-based mean-variance portfolios, minimum-variance portfolios, and strategies for dynamic asset allocation. These policies based their estimates on ten years of past financial data and had to forecast the performance in the following month. For similar results, see Bloomfield et al., 1977. The $1/N$ rule is a version of equal-weight or tallying rules, which have been shown to match or outperform complex weighting policies in speed and accuracy (Czerlinski et al., 1999; Dawes, 1979). See Zweig, 1998, on Markowitz.

10. Ortmann et al., in press; Barber and Odean, 2001.

11. We used the 500 Standard & Poor stocks and 298 German stocks and asked four groups—Chicago pedestrians, University of Chicago students of business, Munich pedestrians, University of Munich students of business—which of the stocks they recognized by name (Borges et al., 1999). We then constructed eight high-recognition portfolios (U.S. and German stocks, for each of the four groups) and evaluated their performance after six months against four benchmarks: market indices, mutual

funds, random ("dartboard") portfolios, and low-recognition portfolios. The high-recognition portfolios outperformed the respective market indices (Dow 30 and Dax 30) and the mutual funds in six out of eight cases and matched or outperformed random portfolios and low-recognition portfolios in all cases. This study prompted a great deal of press coverage and two opposite reactions. There were those, notably financial advisers, who said, "It can't be true," and those who said, "No surprise. We knew it all along." One objection was that we had hit a bull market, yet in two subsequent studies, we hit a bear market and could still replicate the beneficial effect of name recognition (Ortmann et al., in press). However, two other studies did not report name recognition to be of advantage, each of these relying exclusively on the recognition of college students rather then the general public. Boyd, 2001, used students whose recognition of stocks was idiosyncratic and resulted in disproportionate losses or disproportionate gains. Frings et al., 2003, excluded all students who recognized more than 50 percent of the Nemax50 and violated the diversification principle (at least ten stocks in one portfolio) used in our studies. All in all, the studies suggest that portfolios built on collective brand-name recognition perform about as well as and sometimes better than financial experts, the market, and mutual funds do.

12. Sherden, 1998, 107. For instance, between 1968 and 1983, the market outperformed pension fund managers by about .5 percent per year. Adding on management fees, this results in −1 percent per year. In 1995, the Standard & Poor's Index rose by 37 percent, while mutual funds increased by only 30 percent, and the majority (89 percent) could not beat the market. See also Taleb, 2004.

13. Goode, 2001.

14. This magical number was proposed by the psychologist George A. Miller, 1956. Consistent with this number, Malhotra, 1982, concludes that in consumer decisions, ten or more alternatives cause poorer choices.

15. Iyengar and Lepper, 2000.

16. http://www.forbes.com/lists/2003/02/26/billionaireland.html.

17. Lenton et al., 2006.

18. Beilock et al., 2004; Beilock et al., 2002.

19. Johnson and Raab, 2003. The I-shaped bars in Figure 2-3 are standard errors of means.

20. Klein, 1998.
21. Wulf and Prinz, 2001.
22. See Carnap's (1947) "principle of total evidence" and Good's (1967) "total evidence theorem," both of which state that information should never be ignored, and Sober, 1975, for a discussion. Hogarth, in press, reviews four areas in which simple strategies consistently outperform those that use more information: simple actuarial methods predict better than sophisticated clinical judgment; simple methods in time-series forecasting are superior to "theoretically correct" methods; equal weighting (tallying) is often more accurate than using the "optimal" weights; and decisions can often be improved by discarding relevant information. Hogarth concludes that in each of these domains, demonstrations that simple strategies are better than complex ones at predicting complex phenomena have been largely ignored, since the idea is difficult to accept by most researchers. On less is more, see Hertwig and Todd, 2003.

Chapter 3: How Intuition Works

1. Cited in Egidi and Marengo, 2004, 335. Whitehead was an English mathematician and philosopher who coauthored *Principia Mathematica* with Bertrand Russell.
2. Darwin, 1859/1987, 168.
3. The phrase is credited to Jerome Bruner, but the idea is older. For instance, the psychologist Egon Brunswik spoke of vicarious functioning, and Hermann von Helmholtz spoke of unconscious inferences (see Gigerenzer and Murray, 1987).
4. von Helmholtz, 1856–66/1962. In a fascinating series of experiments, Kleffner and Ramachandran, 1992, analyzed in detail how shape is inferred from shading. Bargh, 1989, provided an excellent discussion of automatic processes in general.
5. Baron-Cohen, 1995. Tomasello, 1988, showed that even eighteen-month-old babies use gaze as a cue for reference.
6. Baron-Cohen, 1995, 93.
7. Here and in the following I rely on Sacks, 1995, 259, 270.
8. Rosander and Hofsten, 2002.
9. Barkow et al., 1992; Daly and Wilson, 1988; Pinker, 1997; Tooby and Cosmides, 1992.
10. Cacioppo et al., 2000.

11. For various cognitive theories that investigate the relation between mind and environment, see Anderson and Schooler, 2000, Cosmides and Tooby, 1992, Fiedler and Juslin, 2006, and Gigerenzer, 2000.

12. Axelrod, 1984, used the prisoner's dilemma, a strategic game that has occupied the social sciences for the last half century. Tit for tat was submitted by the psychologist and game theorist Anatol Rapoport. It tends to fail in social environments where people have the option to "quit" in addition to either being kind or being nasty (Delahaye and Mathieu, 1998). Formally, tit for tat involves players making simultaneous moves, but I also use it in a more general sense in which people can also behave successively.

Chapter 4: Evolved Brains

1. Hayek, 1988, 68. The economist and Nobel laureate Friedrich Hayek anticipated several of the ideas I propose, including that behavior is based on rules that typically cannot be verbalized by the acting person, that behavior is contingent on its environment, and that minds do not create institutions but minds and institutions evolve together.

2. Frey and Eichenberger, 1996. The Bush quote is from Todd and Miller, 1999, 287.

3. Richerson and Boyd, 2005, 100. These authors provide a highly recommended introduction to the role of imitation in cultural learning.

4. Tomasello, 1996.

5. Cosmides and Tooby, 1992.

6. Hammerstein, 2003.

7. Freire et al., 2004; Baron-Cohen et al., 1997.

8. Blythe et al., 1999.

9. Turing, 1950, 439. The philosopher Hilary Putnam, 1960, for instance, used Turing's work as a starting point to argue for his distinction between the mind and the brain. Following Putnam, this distinction served to combat attempts at reducing the mind to the physical brain. For many psychologists, it seemed a good basis to establish the autonomy of psychology in relation to neurophysiology and brain sciences.

10. Holland et al., 1986, 2.

11. Silk et al., 2005. The results were independently replicated by Jensen et al., 2006.

12. Thompson et al., 1997.

13. Cameron, 1999.

14. Takezawa et al., 2006.

15. Henrich et al., 2005.

16. Barnes, 1984, vol. 1, 948–49. The argument I sketch is made in more detail by Daston, 1992.

17. Cited in Schiebinger, 1989, 270–72. For Darwin's view, see Darwin, 1874, vol. 2, 316, 326–27.

18. Hall, 1904, 561.

19. The study was conducted by psychologist Richard Wiseman, University of Hertfordshire, during the Edinburgh International Science Festival in 2005. See, for example, the BBC report (http://news.bbc.co.uk/1/hi/uk/4436021.stm).

20. Meyers-Levy, 1989. The ads I mention in the following are featured in her article.

Chapter 5: Adapted Minds

1. Simon, 1990, 7. The British economist Alfred Marshall, 1890/1920, had used this analogy earlier to ridicule the debate between supply-side and demand-side theories, comparing it to an argument about whether the top blade or the bottom blade of a pair of scissors cuts cloth.

2. Simon, 1969/1981, 65. Herbert Simon was a model of a truly interdisciplinary thinker who defies any single classification as psychologist, economist, political scientist, or one of the fathers of artificial intelligence and cognitive science. On the relation between Simon and my work, see Gigerenzer, 2004b.

3. This value results from $.8 \times .8 + .2 \times .2 = .68$. That is, the rat turns left with a probability of .8, and in this case it gets food with a probability of .8, and it turns right with a probability of .2, and it gets food with a probability of .2. On probability matching, see Brunswik, 1939, and Gallistel, 1990.

4. Gigerenzer, 2006.

5. Törngren and Montgomery, 2004.

6. Sherden, 1998, 7.

7. Bröder, 2003; Bröder and Schiffer, 2003; Rieskamp and Hoffrage, 1999.

8. Gigerenzer and Goldstein, 1996; Gigerenzer et al., 1999.

9. Czerlinski et al., 1999. This procedure is known as cross-validation. We repeated it a thousand times in order to average out particular ways of dividing the data. The technical term for the hindsight task is data fitting.

10. In the case of five cities, there are five possible first cities, for each of them four possible second cities, and so on, which results in $5 \times 4 \times 3 \times 2 \times 1$ possible tours. For n cities, this results in $n!$ tours. Some of these tours have the same length; for instance the tour "a, b, c, d, e, and back to a" has the same length as the tour "b, c, d, e, a, and back to b." For five cities, there are five starting points that lead to a tour with the same length, and furthermore there are two directions in which the tour can go. Thus, the number of possible tours needs to be divided by 5×2 to result in $4 \times 3 = 12$ tours with different lengths. In general, this number is $n!/2n = (n-1)!/2$. The campaign tour problem is a version of the traveling salesman problem (Michalewicz and Fogel, 2000, 14).

11. Rapoport, 2003.

12. Michalewicz and Fogel, 2000.

Chapter 6: Why Good Intuitions Shouldn't Be Logical

1. Cartwright, 1999, 1.

2. Tversky and Kahneman, 1982, 98. Note that here and in the following the term *logic* is used to refer to the laws of first-order logic.

3. Gould, 1992, 469. For more on the alleged consequences, see Johnson et al., 1993, Kanwisher, 1989, Stich, 1985.

4. The linguist Paul Grice, 1989, has studied these conversational rules of thumb.

5. Hertwig and Gigerenzer, 1999; see also Fiedler, 1988, Mellers, Hertwig, and Kahneman, 2001. Tversky and Kahneman, 1983, found an effect of a relative frequency formulation in a different problem but stuck with their logical norm (Gigerenzer, 2000). Another reason for this "fallacy" is that people may read "Linda is a bank teller" as implying that "Linda is a bank teller and not active in the feminist movement." This may happen occasionally, but it cannot explain that the "fallacy" largely disappears when the word *probable* is replaced by *how many*.

6. Edwards et al., 2001.

7. Kahneman and Tversky, 1984/2000, 5, 10.

8. Sher and McKenzie, 2006, 467–94.

9. Feynman, 1967, 53.

10. Selten, 1978, 132–33.

11. Wundt, 1912/1973. On artificial intelligence see Copeland, 2004.

12. Gruber and Vonèche, 1977, xxxiv–xxxix.

PART 2: GUT FEELINGS IN ACTION

Chapter 7: Ever Heard Of . . . ?

1. Some have argued that these differences do not reflect different processes but that recognition is just a simpler form of recall (Anderson et al., 1998).

2. Dawkins, 1989, 102.

3. Standing, 1973.

4. Warrington and McCarthy, 1988; Schacter and Tulving, 1994. Laboratory research has demonstrated that memory for mere recognition captures information even in divided-attention learning tasks that are too distracting to allow more substantial memories to be formed (Jacoby et al., 1989).

5. Pachur and Hertwig, 2006. Recognition validities for Wimbledon 2003 Gentlemen's Singles are reported in Serwe and Frings, 2006, and for city populations in Goldstein and Gigerenzer, 2002, and Pohl, 2006. Note that the recognition heuristic is about inferences from memory, not about inferences from givens, where one could be given access to information about the unrecognized alternative.

6. Ayton and Önkal, 2005. Similarly, Andersson et al., 2005, report that laypeople predicted the outcomes of the 2002 Soccer World Cup as well as experts did.

7. Serwe and Frings, 2006. The predictions were made for a sample of 96 matches out of a total of 127 matches. The correlations between the three official ratings ranged from .58 to .74, and the correlation between the two recognition rankings was .64. The betting quotes in Wimbledon, which cannot be compared to the rankings because they are updated after each game, made 79 percent correct predictions. Results were replicated for Wimbledon 2005 by Scheibehenne and Bröder, 2006.

8. Hoffrage, 1995; see also Gigerenzer, 1993.

9. This 60 percent figure is called the *knowledge validity* and defined as the proportion correct when both alternatives are recognized. In contrast, the *recognition validity* is defined as the proportion correct when only one alternative is recognized, and the recognition heuristic is used (Goldstein and Gigerenzer, 2002).

10. The curves in Figure 7-4 can be derived in a formal way. Think about a person who makes predictions about *N* objects, such as tennis players or countries, and who recognizes *n* of these. The number *n* can range be-

tween 0 and N. The task is to predict which of two objects has the higher value on a criterion, such as which player will win the match. There are three possibilities: a person recognizes one of the two objects, none, or both. In the first case, one can use the recognition heuristic, in the second, one has to guess, and in the third, one has to rely on knowledge. The numbers n_{ur}, n_{uu}, and n_{rr} specify how often these cases occur (u = unrecognized, r = recognized). If one relies on the recognition heuristic and every object is paired with every other object once, we then get the following:

Number of correct predictions $= n_{ur}\alpha + n_{uu}1/2 + n_{rr}\beta$.

The first term on the right side of the equation accounts for the correct inferences made by the recognition heuristic; for instance, if there are ten cases where one of the two objects is recognized and the recognition validity α is .80, then one can expect eight correct answers. The second term is for guessing, and the third term equals the number of correct inferences made when knowledge beyond recognition is used (β is the knowledge validity). In general, assuming that a person uses the recognition heuristic and α and β are constant, it can be proven that the recognition heuristic will yield a less-is-more effect if $\alpha > \beta$ (Goldstein and Gigerenzer, 2002).

11. Goldstein and Gigerenzer, 2002.
12. Schooler and Hertwig, 2005.
13. Gigone and Hastie, 1997. The majority rule has been reported for situations in which the correct answer cannot be proved by a group member (as it would be in the case of a math problem).
14. Reimer and Katsikopoulos, 2004. The authors show that the less-is-more effect in groups (reported in the next passage) can be formally derived in the same way, as illustrated by Figure 7-4.
15. Toscani, 1997.
16. Hoyer and Brown, 1990.
17. Allison and Uhl, 1964.
18. Oppenheimer, 2003. See also Pohl, 2006.
19. Volz et al., 2006. In this study, judgments followed the recognition heuristic in 84 percent of the cases, similar to earlier experiments. Moreover, when participants were partially ignorant, that is, they had heard of only one of the two cities, they got more correct answers than when they had heard of both cities.
20. Interview with Simon Rattle (Peitz, 2003).

Chapter 8: One Good Reason Is Enough

1. For experimental evidence that people often rely on only one or a few reasons, see Shepard, 1967, Ford et al., 1989, Shanteau, 1992, Bröder, 2003, Bröder and Schiffer, 2003, and Rieskamp and Hoffrage, 1999.

2. Schlosser, 2002, 50.

3. Dawkins, 1989, 158–61.

4. Cronin, 1991.

5. Gadagkar, 2003.

6. Grafen, 1990, showed that the handicap principle can work both for the evolution of honest signals and in the context of sexual selection.

7. Petrie and Halliday, 1994. The number of eyespots could in turn be inferred from the symmetry of the train; see Gadagkar, 2003.

8. See Sniderman and Theriault, 2004.

9. Menard, 2004.

10. Cited in Neuman, 1986, 174.

11. Also cited in Neuman, 1986, 132.

12. Sniderman, 2000.

13. The string heuristic is a realization of Coombs's (1964) unfolding theory and the concept of proximity voting.

14. Gigerenzer, 1982. This study analyzed voters' reactions to two new parties, the Greens and the European Workers Party (EAP, not reported here). Voters knew very little about the program of the EAP, but their preferences and judgments were as consistent as for the Greens, about which they knew more. There is an important methodological lesson in this (see Gigerenzer, 1982). If I were not interested in how the mind works, I would not ask what cognitive process underlies voters' intuitions, but instead directly take their judgments and analyze them with a standard statistical package. Continuing, I would find and report that the correlations between Left-Right, preferences, and ecological ratings are mostly close to zero. Thus, I would erroneously conclude that preference orders and other issues cannot be explained by Left-Right, and that voters therefore have a differentiated system of reasons in their minds. I might never notice this blunder, cherishing the principle "let's correlate first, and explain later." There are patterns in intuition even when there are none in correlations.

15. Neuman, 1986.

16. Sniderman et al., 1991, 94.

17. Scott, 2002.

18. Bröder, 2000, 2003; Bröder and Schiffer, 2003; Newell et al., 2003. As in the study with parents, these experiments reveal individual differences; that is, people differ on the rules of thumb they use.

19. Keeney and Raiffa, 1993.

20. Tversky and Kahneman, 1982; on conservatism see Edwards, 1968.

21. Todorov, 2003. Bayes's rule is named after the Reverend Thomas Bayes, to whom this rule is attributed. It computes the so-called likelihood ratios for the various possible differences in halftime scores because larger differences should have more weight than smaller ones. Take the Best, by contrast, just looks at who is ahead, and ignores by how much. Although some think of Bayes's rule as the rational way to make decisions in the real world, it is actually impossible to follow this rule in problems of sufficient complexity because it becomes computationally intractable. Bayes's rule can be used when only a few cues are known, but the complex rule is of little use for complex problems.

22. Gröschner and Raab, 2006. In a second study, 208 experts and laypeople were asked to predict the 2002 soccer world champion. Laypeople did significantly better, predicting the winner twice as often as experts. The laypeople more often went with the intuition that the team who had won the most championships beforehand would most likely win again (Brazil), and they were right.

23. When I first reported in a talk that one good reason was better than Franklin's rule, a renowned decision researcher got up and said, "If you want to impress me, you need to show that one good reason can stand up to multiple regression." We took up his challenge and showed for the first time that Take the Best can also outperform it (Gigerenzer and Goldstein, 1996, 1999). We published the data so that everyone could rerun the tests and check the claim. Many who could not believe their eyes confirmed the original result. The next objection was that we had demonstrated this only once. So we extended the test to a total of twenty real-world problems from psychology, economics, biology, sociology, health, and other areas. Multiple regression made an average of 68 percent correct predictions and Take the Best 71 percent. When the news spread in the scientific community, and others independently confirmed the results, the defenders of more-is-always-better suspected the fault no longer in the problems we used, but in how we used multiple

regression. Some experts said that we should have calculated different versions of this method. We did, and found basically the same result (Czerlinski et al., 1999; Martignon and Laskey, 1999). Finally, we were able to prove some of the conditions under which the complex strategy cannot do better than Take the Best (Martignon and Hoffrage, 1999, 2002; Katsikopoulos and Martignon, 2006). That put an end to this objection, but not to the debate. Suddenly multiple regression was no longer the issue; the argument now was that Take the Best needed to be compared with highly complex information-greedy algorithms from artificial intelligence and machine learning. We took up the challenge and found that in many situations, one good reason can predict more accurately than these extremely complex strategies (Brighton, 2006). The complex strategies tested included (1) connectionist models: feed-forward neural networks trained with the back-propagation algorithm, (2) two classic decision tree induction algorithms: classification and regression trees (CART) and C4.5, and (3) exemplar models: the basic nearest neighbor classifier and an elaborate model based on the Nosofsky's GCM model.

24. This is a verbal summary of some of the analytical and simulation results reported in Gigerenzer, Todd, et al., 1999, Katsikopoulos and Martignon, 2006, Martignon and Hoffrage, 2002, Hogarth and Karelaia, 2005a, b, 2006. The crossing of the two lines in Figure 5-2 illustrates the problem of *overfitting*. One can define overfitting in the following way. Consider two random samples from a population (such as two years of temperature measures); the first year is the learning set and the following year is the test set. A model overfits the learning set if an alternative model exists that is less accurate on the learning set but more accurate on the test set.

25. The references for the following examples are in Hutchinson and Gigerenzer, 2005.

26. The original Roman calendar had ten months and the year began with Martius (March); Januarius and Februarius were added later. Julius Caesar changed the start of the year to the first of January, and Quintilis was renamed Julius in his honor; Sextilis later became Augustus in honor of Caesar Augustus (Ifrah, 2000, 7).

Chapter 9: Less Is More in Health Care

1. Naylor, 2001.
2. Berg, Biele, and Gigerenzer, 2007. Informative medical literature, such as the *Guide to Clinical Preventive Services* by the U.S. Preventive Services Task Force, 2002a, is easy to find in a university library or online.
3. Merenstein, 2004.
4. Ransohoff et al., 2002.
5. U.S. Preventive Services Task Force, 2002b.
6. Lapp, 2005.
7. Etzioni et al., 2002.
8. Schwartz et al., 2004.
9. U.S. Food and Drug Administration; see Schwartz et al., 2004.
10. Lee and Brennan, 2002.
11. Gigerenzer, 2002, 93.
12. Domenighetti et al., 1993. Note that Switzerland removed financial barriers to medical care for the whole population almost one century ago. Thus, the rate of treatment (including overtreatment) in the general population is not distorted by uninsured citizens having no access to treatment, as in countries without universal health care (this may explain why a study in the United States did not find different hysterectomy rates for doctors' wives; see Bunker and Brown, 1974).
13. Deveugele et al., 2002; Langewitz, et al. 2002.
14. Kaiser et al., 2004.
15. Wennberg and Wennberg, 1999.
16. Wennberg and Wennberg, 1999, 4.
17. Elwyn et al., 2001.
18. See the reader by Dowie and Elstein, 1988.
19. Elwyn et al., 2001. Concerning our program to improve physicians' and patients' intuitions about risk and uncertainties, see Gigerenzer, 2002, and Hoffrage et al., 2000.
20. Pozen et al., 1984.
21. See Green and Yates, 1995.
22. Green and Mehr, 1997.
23. Corey and Merenstein, 1987; Pearson et al., 1994.
24. For details, see Martignon et al., 2003.
25. Recall that when using the heart disease predictive instrument, the physician calculates a number for each patient, then compares it to a threshold.

If the number is higher than the threshold, the patient is sent into the care unit. This threshold can be set high or low. If it is set high, then fewer patients will be sent into the care unit, and more misses will result. This corresponds to the squares at the left of Figure 9-3. If the threshold is set low, then more people will be sent into the unit, which increases the number of false alarms, as shown by the squares at the right.

26. In a replication study in two other Michigan hospitals, which used the successor of the heart disease predictive instrument, the Acute Coronary Ischemic Time-Insensitive Predictive Instrument (ACI-TIPI), a fast and frugal tree again did as well as the complex method (Green, 1996).

Chapter 10: Moral Behavior

1. Browning, 1998, xvii. I chose this sensitive example because it is one of the best-documented mass murders in history, with the unique feature that the policemen were given the opportunity not to participate in the killing. If you know of other similar examples, please let me know. My short account cannot do justice to the complexity of the situation, and I recommend consulting Browning's book, including the afterword, in which he deals with his critics such as Daniel Goldhagen. Browning (e.g., 209–16) offers a multilayered portrayal of the battalion during their first and subsequent mass killings. The largest group of policemen ended up doing whatever they were asked to, avoiding the risk of confronting authority or appearing to be cowards, yet not volunteering to kill. Increasingly numbed by the violence, they did not think what they were doing was immoral because it was sanctioned by authority. In fact, most did not try to think at all. A second group of "eager" killers who celebrated their murderous deeds increased in numbers over time. The smallest group was the nonshooters, who, with the exception of one lieutenant, however, neither protested against the regime nor reproached their comrades.

2. Browning, 1998, 71.

3. Johnson and Goldstein, 2003. Note that this figure is the proportion of citizens who are potential donors by law, not the actual donation rate. The latter depends on how well the process that matches a donor with a recipient is coordinated and how well the staff is trained—including how fast the donor, typically a traffic accident or stroke victim, is transported to a hospital. In the years 1996 through 2002, by far the best organization

and highest true donation rate was reached in Spain, a country with a presumed consent policy; that is, people are potential donors by default.

4. Johnson and Goldstein, 2003.

5. Johnson et al., 1993.

6. Haidt and Graham, in press, base their five moral dimensions on the work of Shweder et al. (1997), where harm and reciprocity concern the ethics of autonomy, hierarchy and ingroup the ethics of community, and purity the ethics of divinity. The connection of these five dimensions with individual, family, and community (rather than with autonomy, community, and divinity) is not of their doing but my own responsibility. See also Gigerenzer, in press.

7. Kohlberg et al., 1983, 75. In this article, the authors reformulated Kohlberg's (1981) original theory. The following evaluation of the evidence is based on Shweder et al., 1997.

8. Haidt, 2001.

9. Harrison, 1967, 72.

10. Haidt, 2001, 814; see also Nisbett and Wilson, 1977, and Tetlock, 2003.

11. Laland, 2001.

12. Terkel, 1997, 164.

13. The Bail Act 1976 and its subsequent revisions; see Dhami and Ayton, 2001.

14. Dhami and Ayton, 2001, 163. The following quote is from Dhami (August 2003, personal communication).

15. Dhami, 2003.

16. Dhami and Ayton, 2001.

17. Gazzaniga, 1985.

18. Numerous versions of consequentialism exist; see Williams, 1973, and Downie, 1991. Sunstein, 2005, provides an interesting discussion of rules of thumb and consequentialism.

19. Daston, 1988.

20. Bentham, 1789/1907; see Smart, 1967. The hedonic calculus is from Bentham's chapter 4. As requested in his will, Bentham's body was preserved and exhibited in a wooden cabinet at University College London, where, topped with a wax head, it can still be viewed today.

21. Dennett, 1988.

22. Sunstein, 2005; Viscusi, 2000.

Chapter 11: Social Instincts

1. Humphrey, 1976/1988, 19; see also Kummer et al., 1997.

2. Richerson and Boyd, 2005.

3. Cronin, 1991.

4. Darwin, 1874, 178–79.

5. Sober and Wilson, 1998.

6. See Cosmides and Tooby, 1992; Gigerenzer and Hug, 1992.

7. Frevert, 2003.

8. Resche, 2004, 723, 741.

9. Mervyn King, "Reforming the international financial system: The middle way." Speech delivered to a session of the money marketers at the Federal Reserve Bank of New York, September 9, 1999. http://www.bankofengland.co.uk/publications/news/1999/070.htm.

10. Gallup International, 2002.

11. The neurologist Antonio Damasio, 1994, 193–94, reported a patient called Elliot with a damaged frontal lobe. One day, Damasio asked him when the next session should take place and suggested two alternative dates, just a few days apart from each other: "For the better part of a half hour, the patient enumerated reasons for and against each of the two dates: Previous engagements, proximity to other engagements, possible meteorological conditions, virtually anything that one could reasonably think about concerning a simple date. . . . He was walking us through a tiresome cost-benefit analysis, and endless outlining and fruitless comparison of options and possible consequences." When Damasio advised him to take the second date, Elliot simply said, "That's fine."

12. Richerson and Boyd, 2005.

13. For an evolutionary theory of social change, see Boyd and Richerson, 2005.

14. Lightfoot, 2003.

15. Hertle, 1996, 7, 245. The following account is based on Hertle's research.

16. Hertle and Stephan, 1997, 42.

REFERENCES

Allison, R. I., and K. P. Uhl. 1964. Influence of beer brand identification on taste perception. *Journal of Marketing Research* 1: 36–39.

Ambady, N., and R. Rosenthal. 1993. Half a minute: Predicting teacher evaluations from thin slices of nonverbal behavior and physical attractiveness. *Journal of Personality and Social Psychology* 64: 431–41.

Anderson, J. R., D. Bothell, C. Lebiere, and M. Matessa. 1998. An integrated theory of list memory. *Journal of Memory and Language* 38: 341–80.

Anderson, J. R., and L. J. Schooler. 2000. The adaptive nature of memory. In *Handbook of Memory*, ed. E. Tulving and F. I. M. Craik, 557–70. New York: Oxford University Press.

Andersson, P., J. Edman, and M. Ekman. 2005. Predicting the World Cup 2002: Performance and confidence of experts and non-experts. *International Journal of Forecasting* 21: 565–76.

Axelrod, R. 1984. *The Evolution of Cooperation*. New York: Basic Books.

Ayton, P., and D. Önkal. 2005. Effects of ignorance and information on judgments and decisions. Unpublished manuscript.

Babler, T. G., and J. L. Dannemiller. 1993. Role of image acceleration in judging landing location of free-falling projectiles. *Journal of Experimental Psychology: Human Perception and Performance* 19: 15–31.

Barber, B., and T. Odean. 2001. Boys will be boys: Gender, overconfidence, and common stock investment. *Quarterly Journal of Ecomomics* 116: 261–92.

Bargh, J. A. 1989. Conditional automaticity: Varieties of automatic influence in social perception and cognition. In *Unintended Thought*, ed. J. S. Uleman and J. A. Bargh, 3–51. New York: Guilford Press.

Barkow, J. H., L. Cosmides, and J. Tooby, eds. 1992. *The Adapted Mind: Evolutionary Psychology and the Generation of Culture*. New York: Oxford University Press.

Barnes, J., ed. 1984. *The Complete Works of Aristotle*. Princeton, NJ: Princeton University Press.

Baron-Cohen, S. 1995. *Mindblindness: An Essay on Autism and Theory of Mind*. Cambridge, MA: MIT Press.

Baron-Cohen, S., D. Baldwin, and M. Crowson. 1997. Do children with autism use the Speaker's Direction of Gaze (SDG) strategy to crack the code of language? *Child Development* 68: 48–57.

Beilock, S. L., B. I. Bertenthal, A. M. McCoy, and T. H. Carr. 2004. Haste does not always make waste: Expertise, direction of attention, and speed versus accuracy in performing sensorimotor skills. *Psychonomic Bulletin and Review* 11: 373–79.

Beilock, S. L., T. H. Carr, C. MacMahon, and J. L. Starkes. 2002. When paying attention becomes counterproductive: Impact of divided versus skill-focused attention on novice and experienced performance of sensorimotor skills. *Journal of Experimental Psychology: Applied* 8: 6–16.

Bentham, J. 1789. *An Introduction to the Principles of Morals and Legislation*. Oxford, UK: Clarendon Press, 1907.

Berg, N., G. Biele, and G. Gigerenzer. 2007. Logical consistency and accuracy of beliefs: survey evidence on health decision-making among economists. Unpublished manuscript.

Bloomfield, T., R. Leftwich, and J. Long. 1977. Portfolio strategies and performance. *Journal of Financial Economics* 5: 201–18.

Blythe, P. W., P. M. Todd, and G. E. Miller. 1999. How motion reveals intention: Categorizing social interactions. In G. Gigerenzer, P. M. Todd, and the ABC Research Group, *Simple Heuristics That Make Us Smart*, 257–85. New York: Oxford University Press.

Borges, B., D. G. Goldstein, A. Ortmann, and G. Gigerenzer. 1999. Can ignorance beat the stock market? In G. Gigerenzer, P. M. Todd, and the ABC Research Group, *Simple Heuristics That Make Us Smart*, 59–72. New York: Oxford University Press.

Boyd, M. 2001. On ignorance, intuition and investing: A bear market test of the recognition heuristic. *Journal of Psychology and Financial Markets* 2: 150–56.

Boyd, R., and P. J. Richerson. 2005. *The Origin and Evolution of Cultures*. New York: Oxford University Press.

Brighton, H. 2006. Robust inference with simple cognitive models. In *Between a Rock and a Hard Place: Cognitive Science Principles Meet AI-Hard Problems*, 17–22. Papers from the AAAI Spring Syposium (AAAI Technical Report SS-06-03), ed. C. Lebiere and B. Wray. Menlo Park, CA: AAAI Press.

Bröder, A. 2000. Assessing the empirical validity of the "Take-the-Best" heuristic as a model of human probabilistic inference. *Journal of Experimental Psychology: Learning, Memory, and Cognition* 26: 1332–46.

———. 2003. Decision making with the "adaptive toolbox": Influence of environmental structure, intelligence, and working memory load. *Journal of Experimental Psychology* 29: 611–25.

Bröder, A., and S. Schiffer. 2003. Bayesian strategy assessment in multi-attribute decision making. *Journal of Behavioral Decision Making* 16: 193–213.

Browning, C. R. 1998. *Ordinary Men: Reserve Police Battalion 101 and the Final Solution in Poland*. 2nd ed. New York: HarperCollins.

Bruner, J. 1960. *The Process of Education*. Cambridge, MA: Harvard University Press.

Brunswik, E. 1939. Probability as a determiner of rat behavior. *Journal of Experimental Psychology* 25: 175–97.

Bunker, J. P., and B. W. Brown. 1974. The physician-patient as an informed consumer of surgical services. *New England Journal of Medicine* 290: 1051–55.

Bursztajn, H., R. I. Feinbloom, R. M. Hamm, and A. Brodsky. 1990. *Medical Choices, Medical Chances: How Patients, Families, and Physicians Can Cope with Uncertainty*. New York: Routledge, Chapman, & Hall.

Cacioppo, J. T., G. G. Berntson, J. F. Sheridan, and M. K. McClintock. 2000. Multilevel integrative analyses of human behavior: Social neuroscience and the complementing nature of social and biological approaches. *Psychological Bulletin* 126: 829–43.

Cameron, L. A. 1999. Raising the stakes in the Ultimatum Game: Experimental evidence from Indonesia. *Economic Inquiry* 37: 47–59.

Carnap, R. 1947. On the application of inductive logic. *Philosophy and Phenomenological Research* 8: 133–48.

Cartwright, N. 1999. *The Dappled World: A Study of the Boundaries of Science*. Cambridge, UK: Cambridge University Press.

Clark, R. W. 1971. *Einstein: The Life and Times*. New York: The World Publishing Co.

Collett, T. S., and M. F. Land. 1975. Visual control of flight behavior in the hoverfly, Syritta pipiens L. *Journal of Comparative Physiology* 99: 1–66.

Coombs, C. H. 1964. *A Theory of Data*. New York: Wiley.

Copeland, B. J., ed. 2004. *The Essential Turing: Seminal Writings in Computing, Logic, Philosophy, Artificial Intelligence, and Artificial Life, Plus the Secrets of Enigma*. Oxford, UK: Oxford University Press.

Corey, G. A., and J. H. Merenstein. 1987. Applying the acute ischemic heart disease predictive instrument. *Journal of Family Practice* 25: 127–33.

Cosmides, L., and J. Tooby. 1992. Cognitive adaptations for social exchange. In *The Adapted Mind: Evolutionary Psychology and the Generation of Culture*, ed. J. H. Barkow, L. Cosmides, and J. Tooby, 163–228. New York: Oxford University Press.

Cronin, H. 1991. *The Ant and the Peacock: Altruism and Sexual Selection from Darwin to Today*. Cambridge, UK: Cambridge University Press.

Czerlinski, J., G. Gigerenzer, and D. G. Goldstein. 1999. How good are simple heuristics? In G. Gigerenzer, P. M. Todd, and the ABC Reseach Group, *Simple Heuristics That Make Us Smart*, 97–118. New York: Oxford University Press.

Daly, M., and M. Wilson. 1988. *Homicide*. New York: Aldine de Gruyter.

Damasio, A. 1994. *Descartes' Error*. New York: Putnam.

Darwin, C. 1859. *On the Origin of Species*. New York: New York University Press, 1987.

———. 1874. *The Descent of Man, and Selection in Relation to Sex*. 2nd ed. New York: American Home Library.

Daston, L. J. 1988. *Classical Probability in the Enlightenment*. Princeton, NJ: Princeton University Press.

———. 1992. The naturalized female intellect. *Science in Context* 5: 209–35.

Dawes, R. M. 1979. The robust beauty of improper linear models in decision making. *American Psychologist* 34: 571–82.

Dawkins, R. 1989. *The Selfish Gene*. 2nd ed. Oxford, UK: Oxford University Press.

Delahaye, J. P., and P. Mathieu. 1998. Altruismus mit Kündigungsmöglichkeit. *Spektrum der Wissenschaft* (February): 8–14.

DeMiguel, V., L. Garlappi, and R. Uppal. 2006. 1/N. Unpublished manuscript.

Dennett, D. C. 1988. The moral first aid manual. In *The Tanner Lectures on Human Values*. Vol. 8. Ed. S. M. McMurrin, 119–47. Salt Lake City, UT: University of Utah.

Deveugele, M., A. Derese, A. van den Brink-Muinen, J. Bensing, and J. De Maeseneer. 2002. Consultation length in general practice: Cross-sectional study in six European countries. *British Medical Journal* 325: 472–77.

Dhami, M. K. 2003. Psychological models of professional decision-making. *Psychological Science* 14: 175–80.

Dhami, M. K., and P. Ayton. 2001. Bailing and jailing the fast and frugal way. *Journal of Behavioral Decision Making* 14: 141–68.

Dijksterhuis, A., and L. F. Nordgren. 2006. A theory of unconscious thought. *Perspectives on Psychological Science* 1: 95–109.

Domenighetti, G., A. Casabianca, F. Gutzwiller, and S. Martinoli. 1993. Revisiting the most informed consumer of surgical services: The physician-patient. *International Journal of Technology Assessment in Health Care* 9: 505–13.

Dowd, M. 2003. Blanket of dread. *New York Times,* July 30.

Dowie, J., and A. S. Elstein, eds. 1988. *Professional Judgment. A Reader in Clinical Decision Making.* Cambridge, UK: Cambridge University Press.

Downie, R. S. 1991. Moral philosophy. In *The New Palgrave: A Dictionary of Economics.* Vol. 3. Ed. J. Eatwell, M. Milgate, and P. Newman, 551–56. London: Macmillan.

Edwards, A., G. J. Elwyn, J. Covey, E. Mathews, and R. Pill. 2001. Presenting risk information—A review of the effects of "framing" and other manipulations on patient outcomes. *Journal of Health Communication* 6: 61–82.

Edwards, W. 1968. Conservatism in human information processing. In *Formal Representation of Human Judgment,* ed. B. Kleinmuntz, 17–52. New York: Wiley.

Egidi, M., and L. Marengo. 2004. Near-decomposability, organization, and evolution: Some notes on Herbert Simon's contribution. In *Models of a Man: Essays in Memory of Herbert A. Simon,* ed. M. Augier and J. J. March, 335–50. Cambridge, MA: MIT Press.

Einstein, A. 1933. *On the Method of Theoretical Physics.* Oxford: Clarendon Press.

Elman, J. L. 1993. Learning and development in neural networks: The importance of starting small. *Cognition* 48: 71–99.

Elwyn, G. J., A. Edwards, M. Eccles, and D. Rovner. 2001. Decision analysis in patient care. *The Lancet* 358: 571–74.

Etzioni, R., D. F. Penson, J. M. Legler, D. di Tommaso, R. Boer, P. H. Gann, et al. 2002. Overdiagnosis due to prostate-specific antigen screening: Lessons from U.S. prostate cancer incidence trends. *Journal of the National Cancer Institute* 94: 981–90.

Feynman, R. P. 1967. *The Character of Physical Law*. Cambridge, MA: MIT Press.

Fiedler, K. 1988. The dependence of the conjunction fallacy on subtle linguistic factors. *Psychological Research* 50: 123–29.

Fiedler, K., and P. Juslin, eds. 2006. *Information Sampling and Adaptive Cognition*. New York: Cambridge University Press.

Ford, J. K., N. Schmitt, S. L. Schechtman, B. H. Hults, and M. L. Doherty. 1989. Process tracing methods: Contributions, problems, and neglected research questions. *Organizational Behavior and Human Decision Processes* 43: 75–117.

Franklin, B. 1907. Letter to Jonathan Williams (Passy, April 8, 1779). In *The Writings of Benjamin Franklin*. Vol. 7. Ed. A. H. Smyth, 281–82. New York: Macmillan.

Freire, A., M. Eskritt, and K. Lee. 2004. Are eyes windows to a deceiver's soul? Children's use of another's eye gaze cues in a deceptive situation. *Developmental Psychology* 40: 1093–1104.

Frevert, U., ed. 2003. *Vertrauen. Historische Annäherungen*. Göttingen: Vandenhoeck & Ruprecht.

Frey, B. S., and R. Eichenberger. 1996. Marriage paradoxes. *Rationality and Society* 8: 187–206.

Frings, C., H. Holling, and S. Serwe. 2003. Anwendung der Recognition Heuristic auf den Aktienmarkt—Ignorance cannot beat the Nemax50. *Wirtschaftspsychologie* 4: 31–38.

Gadagkar, R. 2003. Is the peacock merely beautiful or also honest? *Current Science* 85: 1012–20.

Gallistel, C. R. 1990. *The Organization of Learning*. Cambridge, MA: MIT Press.

Gallup International. 2002. *Trust Will Be the Challenge of 2003*. Press release, November 8, 2002. http://www.voice-of-the-people.net.

Gazzaniga, M. S. 1985. *The Social Brain: Discovering the Networks of the Mind*. New York: Basic Books.

———. 1998. *The Mind's Past*. Berkeley, CA: University of California Press.

Gigerenzer, G. 1982. Der eindimensionale Wähler. *Zeitschrift für Sozialpsychologie* 13: 217–36.

———. 1993. The bounded rationality of probabilistic mental models. In *Rationality: Psychological and Philosophical Perspectives*, ed. K. I. Manktelow and D. E. Over, 284–313. London: Routledge.

———. 1996. On narrow norms and vague heuristics: A reply to Kahneman and Tversky. *Psychological Review* 103: 592–96.

————. 2000. *Adaptive Thinking: Rationality in the Real World*. New York: Oxford University Press.

————. 2001. Content-blind norms, no norms, or good norms? A reply to Vranas. *Cognition* 81: 93–103.

————. 2002. *Calculated Risks: How to Know When Numbers Deceive You*. New York: Simon & Schuster. (Published in UK as *Reckoning with Risk*, Penguin, 2002).

————. 2004a. Fast and frugal heuristics: The tools of bounded rationality. In *Blackwell Handbook of Judgment and Decision Making*, ed. D. Koehler and N. Harvey, 62–88. Oxford, UK: Blackwell.

————. 2004b. Striking a blow for sanity in theories of rationality. In *Models of a Man: Essays in Honor of Herbert A. Simon*, ed. M. Augier and J. G. March, 389–409. Cambridge, MA: MIT Press.

————. 2006. Follow the leader. *Harvard Business Review* (February): 18.

————. Forthcoming. Moral intuition = Fast and frugal heuristics? In *Moral Psychology: Vol. 2. The Cognitive Science of Morality, Intuition and Diversity*, ed. W. Sinnott-Armstrong. Cambridge, MA: MIT Press.

Gigerenzer, G., and D. G. Goldstein. 1996. Reasoning the fast and frugal way: Models of bounded rationality. *Psychological Review* 103: 650–69.

————. 1999. Betting on one good reason: The Take the Best heuristic. In G. Gigerenzer, P. M. Todd, and the ABC Research Group, *Simple Heuristics That Make Us Smart*, 75–95. New York: Oxford University Press.

Gigerenzer, G., U. Hoffrage, and H. Kleinbölting. 1991. Probabilistic mental models: A Brunswikian theory of confidence. *Psychological Review* 98: 506–28.

Gigerenzer, G., and K. Hug. 1992. Domain-specific reasoning: Social contracts, cheating, and perspective change. *Cognition* 43: 127–71.

Gigerenzer, G., and D. J. Murray. 1987. *Cognition as Intuitive Statistics*. Hillsdale, NJ: Erlbaum.

Gigerenzer, G., and R. Selten, eds. 2001. *Bounded Rationality: The Adaptive Toolbox*. Cambridge, MA: MIT Press.

Gigerenzer, G., P. M. Todd, and the ABC Research Group. 1999. *Simple Heuristics That Make Us Smart*. New York: Oxford University Press.

Gigone, D., and R. Hastie. 1997. The impact of information on small group choice. *Journal of Personality and Social Psychology* 72: 132–40.

Gladwell, M. 2005. *Blink: The Power of Thinking Without Thinking*. New York: Little, Brown.

Goldstein, D. G., and G. Gigerenzer. 2002. Models of ecological rationality: The recognition heuristic. *Psychological Review* 109: 75–90.

Good, I. J. 1967. On the principle of total evidence. *British Journal for the Philosophy of Science* 17: 319–21.

Goode, E. 2001. In weird math of choices, 6 choices can beat 600. *New York Times*, January 9.

Gould, S. J. 1992. *Bully for Brontosaurus: Further Reflections in Natural History*. New York: Penguin Books.

Grafen, A. 1990. Biological signals as handicaps. *Journal of Theoretical Biology* 144: 517–46.

Green, L. A. 1996. Can good enough be as good as the best? Comparative performance of satisficing and optimal decision strategies in chest pain diagnosis. Paper presented at the Society for Medical Decision Making annual meeting, Toronto.

Green, L. A., and D. R. Mehr. 1997. What alters physicians' decisions to admit to the coronary care unit? *The Journal of Family Practice* 45: 219–26.

Green, L. A., and J. F. Yates. 1995. Influence of pseudodiagnostic information on the evaluation of ischemic heart disease. *Annual of Emergency Medicine* 25: 451–57.

Grice, H. P. 1989. *Studies in the Way of Words*. Cambridge, MA: Harvard University Press.

Gröschner, C., and M. Raab. 2006. Vorhersagen im Fussball. Deskriptive und normative Aspekte von Vorhersagemodellen im Sport. *Zeitschrift für Sportpsychologie* 13: 23–36.

Gruber, H. E., and J. J. Vonèche. 1977. *The Essential Piaget*. New York: Basic Books.

Haidt, J. 2001. The emotional dog and its rational tail: A social intuitionist approach to moral judgment. *Psychological Review* 108: 814–34.

Haidt, J., and J. Graham. Forthcoming. When morality opposes justice: Emotions and intuitions related to ingroups, hierarchy, and purity. *Social Justice Research*.

Halberstadt, J., and G. L. Levine. 1999. Effects of reasons analysis on the accuracy of predicting basketball games. *Journal of Applied Social Psychology* 29: 517–30.

Hall, G. S. 1904. *Adolescence*. Vol. 2. New York: Appleton & Co.

Hammerstein, P. 2003. Why is reciprocity so rare in social animals? A

Protestant appeal. In *Genetic and Cultural Evolution of Cooperation*, ed. P. Hammerstein, 83–93. Cambridge, MA: MIT Press.

Harrison, J. 1967. Ethical objectivism. In *The Encyclopedia of Philosophy*. Vol. 3–4. Ed. P. Edwards, 71–75. New York: Macmillan.

Hayek, F. A. 1988. *The Fatal Conceit: The Errors of Socialism*. Chicago: University of Chicago Press.

Henrich, J., R. Boyd, S. Bowles, C. Camerer, E. Fehr, H. Gintis, et al. 2005. "Economic man" in cross-cultural perspective: Behavioral experiments in 15 small-scale societies. *Behavioral and Brain Sciences* 28: 795–855.

Hertle, H.-H. 1996. *Chronik des Mauerfalls: Die dramatischen Ereignisse um den 9. November 1989*. Berlin: Christoph Links Verlag.

Hertle, H.-H., and G.-R. Stephan. 1997. *Das Ende der SED: Die letzten Tage des Zentralkomitees*. Berlin: Christoph Links Verlag.

Hertwig, R., and G. Gigerenzer. 1999. The "conjunction fallacy" revisited: How intelligent inferences look like reasoning errors. *Journal of Behavioral Decision Making* 12: 275–305.

Hertwig, R., and P. M. Todd. 2003. More is not always better: The benefits of cognitive limits. In *The Psychology of Reasoning and Decision Making: A Handbook*, ed. D. Hardman and L. Macchi, 213–31. Chichester, UK: Wiley.

Hoffrage, U. 1995. The adequacy of subjective confidence judgments: Studies concerning the theory of probabilistic mental models. Ph.D. diss., University of Salzburg, Austria.

Hoffrage, U., S. Lindsey, R. Hertwig, and G. Gigerenzer. 2000. Communicating statistical information. *Science* 290: 2261–62.

Hogarth, R. M. 2001. *Educating Intuition*. Chicago: University of Chicago Press.

———. Forthcoming. On ignoring scientific evidence: The bumpy road to enlightenment. In *Ecological Rationality: Intelligence in the World*, ed. P. M. Todd, G. Gigerenzer, and the ABC Research Group. New York: Oxford University Press.

Hogarth, R. M., and N. Karelaia. 2005a. Simple models for multi-attribute choice with many alternatives: When it does and does not pay to face tradeoffs with binary attributes. *Management Science* 51: 1860–72.

———. 2005b. Ignoring information in binary choice with continuous variables: When is less "more"? *Journal of Mathematical Psychology* 49: 115–24.

———. 2006. Regions of rationality: Maps for bounded agents. *Decision Analysis* 3: 124–44.

Holland, J. H., K. J. Holyoak, R. E. Nisbett, and P. R. Thagard. 1986. *Induction: Processes of Inference, Learning and Discovery*. Cambridge, MA: MIT Press.

Horan, D. Forthcoming. Hunches in law enforcement. In *Mere Hunches: Policing in the Age of Terror*, ed. C. Lerner and D. Polsby.

Hoyer, W. D., and S. P. Brown. 1990. Effects of brand awareness on choice for a common, repeat-purchase product. *Journal of Consumer Research* 17: 141–48.

Huberman, G., and W. Jiang. 2006. Offering vs. choice in 401(k) plans: Equity exposure and number of funds. *Journal of Finance.* 61: 763–801.

Humphrey, N. K. 1976. The social function of intellect. In *Machiavellian Intelligence*, ed. R. Byrne and A. Whiten, 13–26. Oxford, UK: Clarendon, 1988.

Hutchinson, J. M. C., and G. Gigerenzer. 2005. Simple heuristics and rules of thumb: Where psychologists and behavioural biologists might meet. *Behavioural Processes* 69: 97–124.

Ifrah, G. 2000. *A Universal History of Numbers*. New York: Wiley.

Iyengar, S. S., and M. R. Lepper. 2000. When choice is demotivating: Can one desire too much of a good thing? *Journal of Personality and Social Psychology* 79: 995–1006.

Jacoby, L. J., V. Woloshyn, and C. Kelley. 1989. Becoming famous without being recognized: Unconscious influences of memory produced by dividing attention. *Journal of Experimental Psychology* 118: 115–25.

James, W. 1890. *The Principles of Psychology*. Vol. 1. Cambridge, MA: Harvard University Press, 1981.

Jensen, K., B. Hare, J. Call, and M. Tomasello. 2006. What's in it for me? Self-regard precludes altruism and spite in chimpanzees. *Proceedings of the Royal Society B: Biological Sciences* 273: 1013–21.

Johnson, E. J., and D. G. Goldstein. 2003. Do defaults save lives? *Science* 302: 1338–39.

Johnson, E. J., J. Hershey, J. Meszaros, and H. Kunreuther. 1993. Framing, probability distortions, and insurance decisions. *Journal of Risk and Uncertainty* 7: 35–51.

Johnson, J. G., and M. Raab. 2003. Take the first: Option generation and resulting choices. *Organizational Behavior and Human Decision Processes* 91: 215–29.

Jones, E. 1953. *The Life and Work of Sigmund Freud*. Vol. 1. New York: Basic Books.

Kahneman, D., P. Slovic, and A. Tversky, eds. 1982. *Judgment Under Uncertainty: Heuristics and Biases*. Cambridge, UK: Cambridge University Press.

Kahneman, D., and A. Tversky. 1996. On the reality of cognitive illusions. *Psychological Review* 103: 582–91.

———. 2000. Choices, values, and frames. In *Choices, Values, and Frames*, ed. D. Kahneman and A. Tversky, 1–16. Cambridge, UK: Cambridge University Press. Reprinted from *American Psychologist* 39 (1984): 341–50.

Kaiser, T., H. Ewers, A. Waltering, D. Beckwermert, C. Jennen, and P. T. Sawicki. 2004. Sind die Aussagen medizinischer Werbeprospekte korrekt? *Arznei-Telegramm* 35: 21–23.

Kanwisher, N. 1989. Cognitive heuristics and American security policy. *Journal of Conflict Resolution* 33: 652–75.

Katsikopoulos, K., and L. Martignon. 2006. Naive heuristics for paired comparisons: Some results on their relative accuracy. *Journal of Mathematical Psychology*, 50, 488–94.

Keeney, R. L., and H. Raiffa. 1993. *Decisions with Multiple Objectives*. Cambridge, UK: Cambridge University Press.

Kleffner, D. A., and V. S. Ramachandran. 1992. On the perception of shape from shading. *Perception and Psychophysics* 52: 18–36.

Klein, G. 1998. *Sources of Power: How People Make Decisions*. Cambridge, MA: MIT Press.

Kohlberg, L. 1981. *The Philosophy of Moral Development: Moral Stages and the Idea of Justice: Vol. 1 of Essays on Moral Development*. San Francisco: Harper and Row.

Kohlberg, L., C. Levine, and A. Hewer. 1983. Moral stages: A current formulation and a response to critics. In *Contributions to Human Development*. Vol. 10. Ed. J. A. Meacham. New York: Karger.

Kummer, H., L. Daston, G. Gigerenzer, and J. Silk. 1997. The social intelligence hypothesis. In *Human by Nature: Between Biology and the Social Sciences*, ed. P. Weingart, P. Richerson, S. D. Mitchell, and S. Maasen, 157–79. Hillsdale, NJ: Erlbaum.

Laland, K. 2001. Imitation, social learning, and preparedness as mechanisms of bounded rationality. In *Bounded Rationality: The Adaptive Toolbox*, ed. G. Gigerenzer and R. Selten, 233–47. Cambridge, MA: MIT Press.

Lanchester, B. S., and R. F. Mark. 1975. Pursuit and prediction in the tracking of moving food by a teleost fish (Acanthaluteres spilomelanurus). *Journal of Experimental Psychology: General* 63: 627–45.

Langewitz, W., M. Denz, A. Keller, A. Kiss, S. Rüttimann, and B. Wössmer. 2002. Spontaneous talking time at start of consultation in outpatient clinic: Cohort study. *British Medical Journal* 325: 682–83.

Lapp, T. 2005. Clinical guidelines in court: It's a tug of war. *American Academy of Family Physicians Report,* February 2005. http://www.aafp.org/x33422.xml (accessed March 1, 2006).

Lee, T. H., and T. A. Brennan. 2002. Direct-to-consumer marketing of high-technology screening tests. *New England Journal of Medicine* 346: 529–31.

Lenton, A. P., B. Fasolo, and P. M. Todd. 2006. When less is more in "shopping" for a mate: Expectations vs. actual preferences in online mate choice. Submitted manuscript.

Lerner, C. 2006. Reasonable suspicion and mere hunches. *Vanderbilt Law Review* 59: 407–74.

Levine, G. M., J. B. Halberstadt, and R. L. Goldstone. 1996. Reasoning and the weighing of attributes in attitude judgements. *Journal of Personality and Social Psychology* 70: 230–40.

Lieberman, M. D. 2000. Intuition: A social cognitive neuroscience approach. *Psychological Bulletin* 126: 109–37.

Lightfoot, L. 2003. Unruly boys taken on pink bus to shame them into behaving. *Daily Telegraph,* May 20.

Luria, A. R. 1968. *The Mind of a Mnemonist.* Cambridge, MA: Harvard University Press.

Malhotra, N. K. 1982. Information load and consumer decision making. *The Journal of Consumer Research* 8: 419–30.

Malkiel, B. G. 1985. *A Random Walk Down Main Street: The Time-Tested Strategy for Successful Investing.* 4th ed. New York: Norton.

Marshall, A. 1890. *Principles of Economics.* 8th ed. London: Macmillan, 1920.

Martignon, L., and U. Hoffrage. 1999. Why does one-reason decision making work? A case study in ecological rationality. In G. Gigerenzer, P. M. Todd, and the ABC Research Group, *Simple Heuristics That Make Us Smart,* 119–40. New York: Oxford University Press.

———. 2002. Fast, frugal and fit: Lexicographic heuristics for paired comparison. *Theory and Decision* 52: 29–71.

Martignon, L., and K. B. Laskey. 1999. Bayesian benchmarks for fast and frugal heuristics. In G. Gigerenzer, P. M. Todd, and the ABC Research Group, *Simple Heuristics That Make Us Smart,* 169–88. New York: Oxford University Press.

Martignon, L., O. Vitouch, M. Takezawa, and M. R. Forster. 2003. Naive and yet enlightened: From natural frequencies to fast and frugal decision trees. In *Thinking: Psychological Perspectives on Reasoning, Judgment and Decision Making*, ed. D. Hardman and L. Macchi, 189–211. Chichester, UK: Wiley.

Martin, R. D. 1994. *The Specialist Chick Sexer: A History, a World View, Future Prospects*. Melbourne: Bernal Publishing.

Matthews, G., R. D. Roberts, and M. Zeidner. 2004. Seven myths about emotional intelligence. *Psychological Inquiry* 15: 179–98.

McBeath, M. K., D. M. Shaffer, and M. K. Kaiser. 1995. How baseball outfielders determine where to run to catch fly balls. *Science* 268: 569–73.

McBeath, M. K., D. M. Shaffer, S. E. Morgan, and T. G. Sugar. 2002. Lack of conscious awareness of how we navigate to catch baseballs. Paper presented at the Toward a Center of Consciousness Conference, University of Arizona, Tucson.

McKenzie, C. R. M., and J. D. Nelson. 2003. What a speaker's choice of frame reveals: Reference points, frame selection, and framing effects. *Psychonomic Bulletin and Review* 10: 596–602.

Mellers, B., R. Hertwig, and D. Kahneman. 2001. Do frequency representations eliminate conjunction effects? An exercise in adversarial collaboration. *Psychological Science* 12: 269–75.

Menard, L. 2004. The unpolitical animal. *The New Yorker*, August 30.

Merenstein, D. 2004. Winners and losers. *Journal of the American Medical Association* 7: 15–16.

Meyers-Levy, J. 1989. Gender differences in information processing: A selectivity interpretation. In *Cognitive and Affective Responses to Advertising*, ed. P. Cafferata and A. Tybout, 219–60. Lexington, MA: Lexington Books.

Michalewicz, Z., and D. Fogel. 2000. *How to Solve It: Modern Heuristics*. New York: Springer.

Miller, G. A. 1956. The magical number seven, plus or minus two: Some limits on our capacity of processing information. *Psychological Review* 63: 81–97.

Myers, D. G. 2002. *Intuition: Its Powers and Perils*. New Haven, CT: Yale University Press.

Naylor, C. D. 2001. Clinical decisions: From art to science and back again. *The Lancet* 358: 523–24.

Neuman, W. R. 1986. *The Paradox of Mass Politics*. Cambridge, MA: Harvard University Press.

Newell, B. R., N. Weston, and D. R. Shanks. 2003. Empirical tests of a fast and frugal heuristic: Not everyone "takes-the-best." *Organizational Behavior and Human Decision Processes* 91: 82–96.

Newport, E. L. 1990. Maturational constraints on language learning. *Cognitive Science* 14: 11–28.

Nisbett, R. E., and T. D. Wilson. 1977. Telling more than we can know: Verbal reports on mental processes. *Psychological Review* 84: 231–59.

Oppenheimer, D. M. 2003. Not so fast! (and not so frugal!): Rethinking the recognition heuristic. *Cognition* 90: B1–B9.

Ortmann, A., G. Gigerenzer, B. Borges, and D. G. Goldstein. Forthcoming. The recognition heuristic: A fast and frugal way to investment choice? In *Handbook of Experimental Economics Results,* ed. C. R. Plott and V. L. Smith. Amsterdam: Elsevier/North-Holland.

Pachur, T., and R. Hertwig. 2006. On the psychology of the recognition heuristic: Retrieval primacy as a key determinant of its use. *Journal of Experimental Psychology: Learning, Memory, and Cognition* 32: 983–1002.

Payne, J. W., J. R. Bettman, and E. J. Johnson. 1993. *The Adaptive Decision Maker.* Cambridge, UK: Cambridge University Press.

Pearson, S. D., L. Goldman, T. B. Garcia, E. F. Cook, and T. H. Lee. 1994. Physician response to a prediction rule for the triage of emergency department patients with chest pain. *Journal of General Internal Medicine* 9: 241–47.

Peitz, C. 2003. Interview with Simon Rattle: Das Leben ist keine Probe [Life is not a rehearsal]. *Der Tagesspiegel* 23 (December 29, 2003). http://archiv.tagesspiegel.de/archiv/2029.2012.2003/909678.asp (accessed March 21, 2006).

Petrie, M., and T. Halliday. 1994. Experimental and natural changes in the peacock's (*Pavo cristatus*) train can affect mating success. *Behavioral and Ecological Sociobiology* 35: 213–17.

Pinker, S. 1997. *How the Mind Works.* New York: Norton.

Pohl, R. F. 2006. Empirical tests of the recognition heuristic. *Journal of Behavioral Decision Making* 19: 251–71.

Polya, G. 1954. *Mathematics and Plausible Reasoning.* Vol. 1. Princeton, NJ: Princeton University Press.

Pozen, M. W., R. B. D'Agostino, H. P. Selker, P. A. Sytkowski, and W. B. Hood. 1984. A predictive instrument to improve coronary-care-unit admission practices in acute ischemic heart disease. *New England Journal of Medicine* 310: 1273–78.

Putnam, H. 1960. Minds and machines. In *Dimensions of Mind*, ed. S. Hook, 138–64. New York: New York University Press.

Ransohoff, D. F., M. McNaughton Collins, and F. J. Fowler Jr. 2002. Why is prostate cancer screening so common when the evidence is so uncertain? A system without negative feedback. *The American Journal of Medicine* 113: 663–67.

Rapoport, A. 2003. Chance, utility, rationality, strategy, equilibrium. *Behavioral and Brain Sciences* 26: 172–73.

Reimer, T., and K. Katsikopoulos. 2004. The use of recognition in group decision-making. *Cognitive Science* 28: 1009–29.

Resche, C. 2004. Investigating "Greenspanese": From hedging to "fuzzy transparency." *Discourse and Society* 15: 723–44.

Richerson, P. J., and R. Boyd. 2005. *Not by Genes Alone*. Chicago: University of Chicago Press.

Rieskamp, J., and U. Hoffrage. 1999. When do people use simple heuristics and how can we tell? In G. Gigerenzer, P. M. Todd, and the ABC Research Group, *Simple Heuristics That Make Us Smart*, 141–67. New York: University of Oxford Press.

Rosander, K., and C. Hofsten. 2002. Development of gaze tracking of small and large objects. *Experimental Brain Research* 146: 257–64.

Sacks, O. 1995. *An Anthropologist from Mars*. New York: Vintage Books.

Saxberg, B. V. H. 1987. Projected free fall trajectories: I. Theory and simulation. *Biological Cybernetics* 56: 159–75.

Schacter, D. L. 2001. *The Seven Sins of Memory: How the Mind Forgets and Remembers*. New York: Houghton Mifflin.

Schacter, D. L., and E. Tulving. 1994. What are the memory systems of 1994? In *Memory Systems, 1994*, ed. D. L. Schacter and E. Tulving, 1–38. Cambridge, MA: MIT Press.

Scheibehenne, B., and A. Bröder. 2006. Predicting Wimbledon tennis results 2005 by mere player name recognition. Unpublished manuscript.

Schiebinger, L. 1989. *The Mind Has No Sex? Women in the Origins of Modern Science*. Cambridge, MA: Harvard University Press.

Schlosser, E. 2002. *Fast Food Nation*. London: Penguin.

Schooler, L. J., and J. R. Anderson. 1997. The role of process in the rational analysis of memory. *Cognitive Psychology* 32: 219–50.

Schooler, L. J., and R. Hertwig. 2005. How forgetting aids heuristic inference. *Psychological Review* 112: 610–28.

Schwartz, B., A. Ward, J. Monterosso, S. Lyubomirsky, K. White, and D. R. Lehman. 2002. Maximizing versus satisficing: Happiness is a matter of choice. *Journal of Personality and Social Psychology* 83: 1178–97.

Schwartz, L. M., S. Woloshin, F. J. Fowler Jr., and H. G. Welch. 2004. Enthusiasm for cancer screening in the United States. *Journal of the American Medical Association* 291: 71–78.

Scott, A. 2002. Identifying and analyzing dominant preferences in discrete choice experiments: An application in health care. *Journal of Economic Psychology* 23: 383–98.

Selten, R. 1978. The chain-store paradox. *Theory and Decision* 9: 127–59.

Serwe, S., and C. Frings. 2006. Who will win Wimbledon 2003? The recognition heuristic in predicting sports events. *Journal of Behavioral Decision Making* 19: 321–32.

Shaffer, D. M., S. M. Krauchunas, M. Eddy, and M. K. McBeath. 2004. How dogs navigate to catch Frisbees. *Psychological Science* 15: 437–41.

Shaffer, D. M., and M. K. McBeath. 2005. Naive beliefs in baseball: Systematic distortion in perceived time of apex for fly balls. *Journal of Experimental Psychology: Learning, Memory, and Cognition* 31: 1492–1501.

Shanteau, J. 1992. How much information does an expert use? Is it relevant? *Acta Psychologica* 81: 75–86.

Shanks, D. R. 2005. Implicit learning. In *Handbook of Cognition*, ed. K. Lamberts and R. Goldstone, 202–20. London: Sage.

Shepard, R. N. 1967. On subjectively optimum selections among multiattribute alternatives. In *Decision Making*, ed. W. Edwards and A. Tversky, 257–83. Baltimore, MD: Penguin Books.

Sher, S., and C. R. M. McKenzie. 2006. Information leakage from logically equivalent frames. *Cognition* 101: 467–94.

Sherden, W. A. 1998. *The Fortune Sellers: The Big Business of Buying and Selling Predictions*. New York: Wiley.

Shweder, R. A., N. C. Much, M. Mahaptra, and L. Park. 1997. The "big three" of morality (autonomy, community, and divinity), and the "big three" explanations of suffering, as well. In *Morality and Health*, ed. A. Brandt and P. Rozin, 119–69. New York: Routledge.

Silk, J. B., S. F. Brosnan, J. Vonk, J. Henrich, D. J. Povinelli, A. S. Richardson, et al. 2005. Chimpanzees are indifferent to the welfare of unrelated group members. *Nature* 437: 1357–59.

Simon, H. A. 1969. *The Sciences of the Artificial*. 2nd ed. Cambridge, MA: MIT Press, 1981.

——. 1990. Invariants of human behavior. *Annual Review of Psychology* 41: 1–19.

——. 1992. What is an "explanation" of behavior? *Psychological Science* 3: 150–61.

Smart, J. J. C. 1967. Utilitarianism. In *The Encyclopedia of Philosophy*. Vol. 8. Ed. P. Edwards, 206–12. New York: Macmillan.

Sniderman, P. M. 2000. Taking sides: A fixed choice theory of political reasoning. In *Elements of Reason: Cognition, Choice, and the Bounds of Rationality*, ed. A. Lupia, M. D. McCubbins, and S. L. Popkin, 74–84. New York: Cambridge University Press.

Sniderman, P. M., R. A. Brody, and P. E. Tetlock. 1991. *Reasoning and Choice: Explorations in Political Psychology*. New York: Cambridge University Press.

Sniderman, P. M., and S. M. Theriault. 2004. The dynamics of political argument and the logic of issue framing. In *Studies in Public Opinion: Gauging Attitudes, Nonattitudes, Measurement Error and Change*, ed. W. E. Saris and P. M. Sniderman, 133–65. Princeton, NJ: Princeton University Press.

Sober, E. 1975. *Simplicity*. Oxford, UK: Oxford University Press.

Sober, E., and D. S. Wilson. 1998. *Unto Others: The Evolution and Psychology of Unselfish Behavior*. Cambridge, MA: Harvard University Press.

Standing, L. 1973. Learning 10,000 pictures. *Quarterly Journal of Experimental Psychology* 25: 207–22.

Stich, S. P. 1985. Could man be an irrational animal? *Synthese* 64: 115–35.

Sunstein, C. R. 2005. Moral heuristics. *Behavioral and Brain Sciences* 28: 531–42.

Takezawa, M., M. Gummerum, and M. Keller. 2006. A stage for the rational tail of the emotional dog: Roles of moral reasoning in group decision making. *Journal of Economic Psychology* 27: 117–39.

Taleb, N. N. 2004. *Fooled by Randomness: The Hidden Role of Chance in Life and in the Markets*. New York: Texere.

Terkel, S. 1997. *My American Century*. New York: The New Press.

Tetlock, P. E. 2003. Thinking the unthinkable: Sacred values and taboo cognitions. *Trends in Cognitive Sciences* 7: 320–24.

Thompson, C., J. Barresi, and C. Moore. 1997. The development of future-oriented prudence and altruism in preschool children. *Cognitive Development* 12: 199–212.

Todd, J. T. 1981. Visual information about moving objects. *Journal of Experimental Psychology: Human Perception and Performance* 7: 8795–8810.

Todd, P. M., and G. Gigerenzer. 2003. Bounding rationality to the world. *Journal of Economic Psychology* 24: 143–65.

Todd, P. M., and G. F. Miller. 1999. From pride and prejudice to persuasion: Satisficing in mate search. In G. Gigerenzer, P. M. Todd, and the ABC Research Group, *Simple Heuristics That Make Us Smart*, 287–308. New York: Oxford University Press.

Todorov, A. 2003. Predicting real outcomes: When heuristics are as smart as statistical models. Unpublished manuscript.

Tomasello, M. 1988. The role of joint-attentional processes in early language acquisition. *Language Sciences* 10: 69–88.

———. 1996. Do apes ape? In *Social Learning in Animals: The Roots of Culture*, ed. B. G. Galef Jr. and C. M. Heyes, 319–46. New York: Academic Press.

Tooby, J., and L. Cosmides. 1992. The psychological foundations of culture. In *The Adapted Mind: Evolutionary Psychology and the Generation of Culture*, ed. J. Barkow, L. Cosmides, and J. Tooby, 19–136. New York: Oxford University Press.

Törngren, G., and H. Montgomery. 2004. Worse than chance? Performance and confidence among professionals and laypeople in the stock market. *Journal of Behavioral Finance* 5: 148–53.

Toscani, O. 1997. *Die Werbung ist ein lächelndes Aas*. Frankfurt a.M.: Fischer.

Turing, A. M. 1950. Computing machinery and intelligence. *Mind* 59: 433–60.

Tversky, A., and D. Kahneman. 1982. Judgments of and by representativeness. In *Judgment Under Uncertainty: Heuristics and Biases*, ed. D. Kahneman, P. Slovic, and A. Tversky, 84–98. Cambridge, UK: Cambridge University Press.

———. 1983. Extensional versus intuitive reasoning: The conjunction fallacy in probability judgment. *Psychological Review* 90: 293–315.

U.S. Preventive Services Task Force. 2002a. *Guide to Clinical Preventive Services: Report of the U.S. Preventive Services Task Force*. 3rd ed. Baltimore, MD: Williams & Wilkins.

———. 2002b. Screening for prostate cancer: Recommendations and rationale. *Annals of Internal Medicine* 137: 915–16.

Viscusi, W. K. 2000. Corporate risk analysis: A reckless act? *Stanford Law Review* 52: 547–97.

Volz, K. G., L. J. Schooler, R. I. Schubotz, M. Raab, G. Gigerenzer, and D. Y. von Cramon. 2006. Why you think Milan is larger than Modena: Neural correlates of the recognition heuristic. *Journal of Cognitive Neuroscience* 18: 1924–36.

von Helmholtz, H. 1856–66. *Treatise on Psychological Optics,* trans. J. P. C. Southall. New York: Dover, 1962.

Vranas, P. B. M. 2001. Single-case probabilities and content-neutral norms: A reply to Gigerenzer. *Cognition* 81: 105–11.

Warrington, E. K., and R. A. McCarthy. 1988. The fractionation of retrograde amnesia. *Brain and Cognition* 7: 184–200.

Wegner, D. M. 2002. *The Illusion of Conscious Will.* Cambridge, MA: MIT Press.

Wennberg, J., and D. Wennberg, eds. 1999. *Dartmouth Atlas of Health Care in Michigan.* Chicago: AHA Press.

Williams, B. 1973. A critique of utilitarianism. In *Utilitarianism: For and Against,* ed J. J. C. Smart and B. Williams, 77–150. Cambridge, UK: Cambridge University Press.

Wilson, T. D. 2002. *Strangers to Ourselves: Discovering the Adaptive Unconscious.* Cambridge, MA: Harvard University Press.

Wilson, T. D., D. J. Lisle, J. W. Schooler, S. D. Hodges, K. J. Klaaren, and S. J. LaFleur. 1993. Introspecting about reasons can reduce post-choice satisfaction. *Personality and Social Psychology Bulletin* 19: 331–39.

Wilson, T. D., and J. W. Schooler. 1991. Thinking too much: Introspection can reduce the quality of preferences and decisions. *Journal of Personality and Social Psychology* 60: 181–92.

Wulf, G., and W. Prinz. 2001. Directing attention to movement effects enhances learning: A review. *Psychonomic Bulletin and Review* 8: 648–60.

Wundt, W. 1912. *An Introduction to Psychology,* trans. R. Pintner. New York: Arno, 1973.

Zajonc, R. B. 1980. Feeling and thinking: Preferences need no inferences. *American Psychologist* 35: 151–75.

Zweig, J. 1998. Five investing lessons from America's top pension fund. *Money,* 115–18.

INDEX